DATE DUE

NO 6 '96		
NO 1 '97		
SE 21 '98		
NO 23 '98		
MR 1 '99		
MR 2 '99		
AP 5 '99		
SE 24 '99		
MR 15 '10		

DEMCO 38-296

Oceanography

Sixth Edition

M. GRANT GROSS

Merrill, an imprint of
Macmillan Publishing Company
New York

Collier Macmillan Canada, Inc.
Toronto

Maxwell Macmillan International Publishing Group
New York   Oxford   Singapore   Sydney

Cover Photo: Frank S. Balthis

Macmillan Publishing Company
866 Third Avenue
New York, NY 10022

This book was set in Meridien.

Administrative Editor: Stephen Helba
Production Editor: Ben Ko
Art Coordinator: Lorraine Woost
Cover Designer: Brian Deep

Library of Congress Catalog Card Number: 89–85101
International Standard Book Number: 0–675–21278–2
Printed in the United States of America
PRINTING   4 5 6 7 8 9 YEAR   3 4 5

Other Books in the Merrill Earth Science Series

Preface

Oceanography is the scientific study of the sea. It is a young science which began with the exploring voyage of the Challenger Expedition (1872–1876), and in the past few decades it has changed the way we view the Earth. Since the 1970s, our added capability of seeing the Earth and ocean from space has helped revolutionize our ideas about the ocean and atmosphere. This edition builds on those revolutions.

Advances in oceanography continue at a rapid pace. This sixth edition of *Oceanography* includes the most recent discoveries on our changing world and its climate. It also retains the focus of covering the basic concepts about how the ocean works and how it affects our lives.

The book is still built on high-school-level mathematics and a general science background. It is intended for use in a one-quarter course or for a one-semester course with supplemental materials. It could also be a supplemental text in subjects such as marine geology or marine biology.

A number of new illustrations, based on suggestions and advice from users of previous editions, are interspersed throughout the text. Helpful comments during this edition's review process came from Robert Feller, University of South Carolina; Stanley Jacobs, University of Michigan; and Dennis Kelly, Orange Coast College. As with this book's preceding versions, the attempt is to pack a story about the world ocean into a tiny compass, hoping it will introduce readers to a deeper understanding of how the ocean works.

Contents

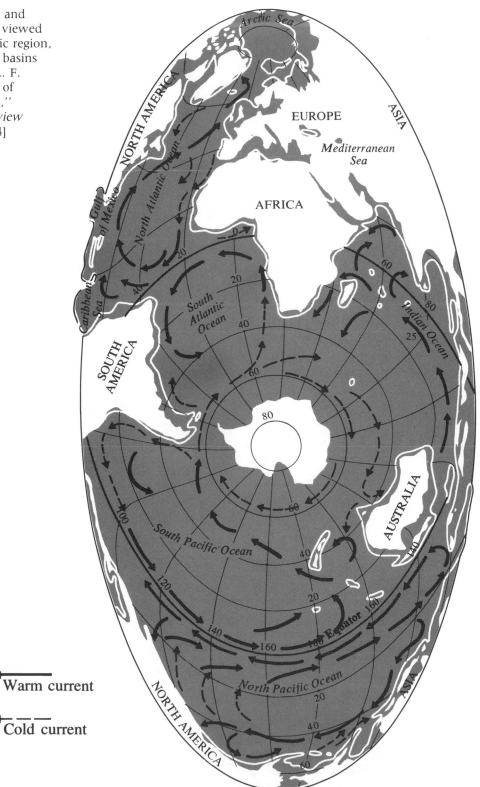

The world ocean and surface currents, viewed from the Antarctic region, where the ocean basins connect. [After A. F. Spilhaus, "Maps of the World Ocean," *Geographical Review* 32 (1942), p. 434]

1
Earth—the Water Planet

Earth is the water planet. Since Earth solidified more than 4 billion years ago, volcanoes have extracted water from rocks deep in its interior, and through volcanic eruptions, released the water to collect in ocean basins (Figure 1.1). Nearly 71% of the Earth's surface is now covered by ocean waters, averaging 3730 m (12,200 ft) deep. At any instant only a small fraction of the water—about 2%—is in the atmosphere or in lakes, rivers, or glaciers. Some of it remains in rocks but does not participate in the day-to-day cycling at the planet's surface. For the most part, nearly all of the water remains in the ocean.

Most people live within a few hundred kilometers of the sea. All of us live within the ocean's influence. The ocean provides us with recreation and food, receives our wastes, and serves as our global highway. It is the source from which the atmosphere draws its water, and it stores and later releases much of the solar energy that powers the winds and causes our weather. By making a simple contrast between the large daily temperature changes in an inland desert and the limited temperature range in a coastal climate, we can see how deeply the ocean controls our lives.

In short, ours is a water-conditioned existence. By studying the ocean, we learn not only about how much of life rose from the ocean, but about how much of life continues to depend on it.

FIGURE 1.1
Earth as seen on August 8, 1980, from a satellite above South America. North America
is left center. The white areas are clouds; spiral ones are hurricanes. Hurricane Allen
is in the Gulf of Mexico. Hurricane Isis is in the Pacific, west of Mexico. (Photograph
courtesy NASA)

1.1   AGE AND ORIGIN OF THE OCEAN

The ocean is an ancient feature of the Earth's surface, and through time the
records of its origins have been obscured. We know that ocean basins and continents were formed early. (We discuss these processes in Chapter 2.) We also
know that nearly 4 billion years ago (Figure 1.2), water that was originally

bound in interior rocks began accumulating at the Earth's surface because of volcanic action. We have evidence for this in ancient rocks—approximately 3.8 billion years old—that contain waterworn pebbles and other features that suggest they were deposited in water. Some of these ancient rocks (3.4 billion years old) contain primitive one-celled bacterialike organisms, visible with electron microscopes.

We know little about changes in seawater composition through time. It appears that the composition of sea salt has changed little. Oxygen has probably become more abundant in the atmosphere, and consequently in the ocean, due to **photosynthesis,** which is the formation of new plant material using energy from sunlight and carbon dioxide plus water.

1.2   SHORELINES AND SEA LEVEL

Shorelines—the boundaries between land and water—continually change, although very slowly. They move as continents rise and fall. Slow vertical movements of continents have occurred repeatedly during Earth's history and continue today. The dramatic effects of their movement can be seen in the elevated ancient beaches in California and in the sunken Greek and Roman temples around the Mediterranean.

Shorelines also move as sea level rises and falls. Changes in sea level occur because of fluctuations in the amount of seawater in the ocean, which is influenced by a number of factors, including river drainage and ice formation. Large changes in sea level accompany the growth or disappearance of large ice sheets on land, called **glaciers.** As glaciers form they take water out of the ocean and as they melt the water is returned (Figure 1.3).

Over the past 3 million years the Earth has experienced many glacial advances and retreats. About 20,000 years ago, large areas of the northern continents were ice-covered and the sea stood about 130 m (430 ft) below its present level (Figure 1.3). Much of the now submerged edge of continents was then dry land. Asia and North America, for example, were connected by land called the Bering Land Bridge, which is now covered by shallow waters. Early humans migrated across this area from Asia to populate North and South America, and mammoths (extinct elephant-like animals) grazed on the submerged continental margins. Rivers flowed through large valleys on lands now submerged.

As the glaciers melted, sea level rose (Figure 1.3), flooding river valleys. On glacier-carved mountainous coasts, steep-sided inlets, called **fjords,** were flooded. If all the water still in the Antarctic and Greenland ice sheets were returned to the ocean, sea level would rise about 50 m (160 ft), flooding large portions of low-lying coastal plains and many coastal cities—up to twenty floors in downtown New York City.

Even slower changes in shorelines occur as continents and ocean basins slowly move, driven by forces acting below the Earth's surface. This process is discussed in Chapter 2.

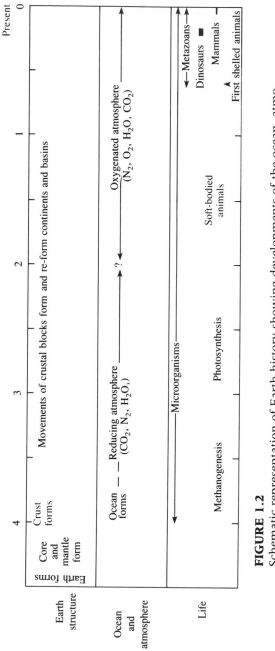

FIGURE 1.2

Schematic representation of Earth history showing developments of the ocean, atmosphere, and life.

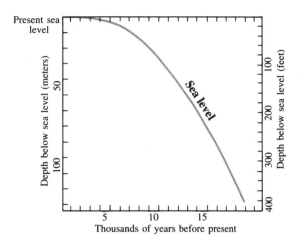

FIGURE 1.3
Sea-level changes during the past 20,000 years. These changes were caused by the melting and retreat of continental glaciers during the latest phase of the Ice Age.

1.3 DISTRIBUTION OF LAND AND WATER

The ocean covers 70.8% of the Earth's surface (Figure 1.4). Although the average depth of the ocean is 3730 m (2.32 statute miles), its depth is insignificant (1/1700) when compared to the Earth's radius (6378 km). (If the Earth were a basketball, the ocean would be only a thin film of water on it.) Excluding fluids in porous, nearsurface rocks, the ocean contains 98% of the planet's free water (Figure 1.5).

Ocean basins and continents are unevenly distributed over the Earth's surface (Figure 1.6). Continents generally have an ocean opposite them on the other side of the Earth. Most land (67%) lies in the Northern Hemisphere; the Southern Hemisphere is mainly water.

The ocean covers 81% of the Earth's surface between latitudes 40° and 60°S, and in the "Roaring Forties" there is no land to impede winds or ocean currents. Looking down on the South Pole from space, *the world ocean is a broad band surrounding Antarctica with three northward projecting gulfs.* (See the map on page xi to verify this for yourself.) *There is only one interconnected ocean,* but for convenience we divide it into three parts: the Atlantic (including the Arctic), the Indian, and the Pacific (Figure 1.6). Where land boundaries are absent, we delineate the oceans using north-south lines. For example, we separate the Pacific and Indian oceans by a line from along 150°E from Indonesia and Australia to Antarctica. We use the islands of Indonesia, north of Australia, as a natural boundary between the Pacific and Indian oceans, and the shallow Bering Strait (because it blocks all but shallow ocean currents) to separate the Pacific Ocean and the Arctic Sea. A line between

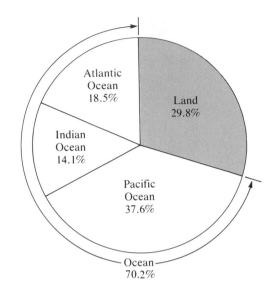

FIGURE 1.4
Ocean covers 70.2% of the Earth's surface. The Pacific is as large as the Atlantic and
Indian oceans combined.

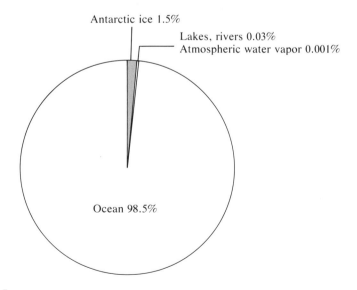

FIGURE 1.5
The ocean contains most of the free water on the Earth's surface (excluding water in
rocks).

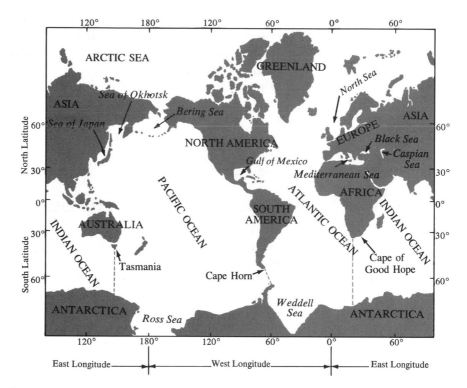

FIGURE 1.6
Continents and ocean basins, showing the boundaries of the major ocean basins.

Cape Horn (the southern tip of South America) and Antarctica, on longitude 60°W, separates the Pacific and Atlantic.

The **Pacific Ocean,** largest of the ocean basins, is nearly as large as the Indian and Atlantic oceans combined, containing slightly more than half the water in the world's ocean. The Pacific is the deepest basin because it includes relatively few shallow seas and has many deep trenches. These trenches are associated with active volcanoes and frequent earthquakes and form the Pacific "Rim of Fire," a belt of mountain building.

Few large rivers discharge into the Pacific Ocean. Its surface area is ten times larger than that of the surrounding lands that drain into it, and because of its size, it is less affected by the presence of continents than are the other ocean basins. Conditions in the Pacific—especially in the South Pacific—most nearly resemble the conditions on a water-covered Earth.

The **Atlantic Ocean** is a relatively narrow, twisted body of water, bounded by roughly parallel continental margins. Including the Arctic Sea, the Atlantic Ocean has the greatest north-south extent of all the ocean basins, connecting the polar regions of the north and south. Because of its many shallow, marginal seas, wide continental shelves, and the Mid-Atlantic Ridge, the Atlantic is the shallowest of the three ocean basins. Many large rivers, including the Amazon and the Congo (the world's largest rivers), discharge into it, making it only 1.6 times larger than its drainage area.

We separate the Indian and Atlantic oceans by a north-south line between the Cape of Good Hope and Antarctica along longitude 20°E (Figure 1.6). The **Indian Ocean** extends only a short distance into the Northern Hemisphere; its northern limit is about 25°N. The northern portion of the Indian Ocean is strongly influenced by the discharges of large rivers (Ganges and Brahmaputra) and the seasonally changeable monsoon winds. (We discuss them in Chapter 5.)

1.4   GENERAL CHARACTERISTICS OF THE OCEAN

Since the ocean basins are interconnected, processes acting in the most remote basin eventually affect the entire ocean. One illustration of this is the relationship between the Mediterranean and the Atlantic. The arid climate around the Mediterranean Sea causes extensive evaporation, making its waters saltier than the adjacent Atlantic. Because of the interconnection between basins, though, the warm, salty waters from the Mediterranean enter the Atlantic Ocean along the bottom of the Strait of Gibraltar (they can be detected about 1.5 km—nearly 1 mi—below the surface across much of the North Atlantic). This injection of salty waters makes the North Atlantic the saltiest ocean.

Because of the ocean's age (more than 4,000 million years) and slow rate of change, the chemical constituents in its sea water are well mixed. Bottom waters in the deep-ocean basins return to the surface in 500 to 1,000 years: thus, ocean waters are thoroughly mixed over billions of years, and seawaters become nearly indistinguishable chemically, regardless of where they are taken. (As we shall see, the organisms in the waters differ greatly depending on the time and place of sampling.)

1.5   ELEVATIONS AND DEPRESSIONS

Elevations on land and on the bottom of the ocean basins have two dominant levels. One, averaging around 840 m (2750 ft) above sea level, is the continental surface. The other, averaging 3740 m (2.32 miles) below sea level, is deep-ocean bottom (Figure 1.7).

These two levels of the Earth's surface reflect fundamental differences in composition and density between continents and ocean basins. *Continents consist of granitic rocks rich in silica and aluminum, with densities (mass per unit volume) of about 2.8 grams per cubic centimeter (g/cm^3). Ocean basins are underlain by basaltic rocks rich in iron and magnesium, with an average density of about 3.0 g/cm^3. The Earth's crust floats on the underlying, nearly liquid mantle, which has an average density of about 4.5 g/cm^3.* Continents float higher

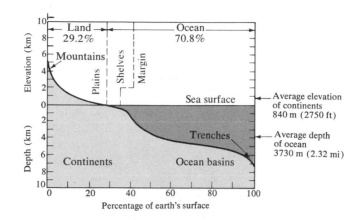

FIGURE 1.7

Hypsometric curve showing the percentage of the Earth's surface above any given depth or elevation. [After H.U. Sverdrup, M.W. Johnson, and R.H. Fleming. *The Oceans: Their Physics, Chemistry, and General Biology*, © 1970, p. 19] Reprinted by permission of Prentice-Hall, Inc., Englewood Cliffs, NJ.

than ocean basins because granitic rocks are less dense than basaltic rocks. At depths less than 2000 m, the ocean bottom is usually continental in composition. At depths greater than 4000 m, the ocean bottom is typically oceanic crust.

Mountains higher than 6000 m (20,000 ft) are rare (Figure 1.5). So are parts of the ocean floor deeper than 6000 m. These extreme heights and depths are less than 1% of the Earth's surface.

1.6   TECHNIQUES FOR STUDYING THE OCEAN

Instruments lowered from ships (Figure 1.8) are the principal means of studying ocean features and processes. Until about 1950, large oceanographic expeditions were usually mounted by a single country, as was the nineteenth-century *Challenger* Expedition conducted by Great Britain. The enormous areas of the ocean and the increasing need for more observations led to large expeditions involving ships of many nations working together. Such expeditions were expensive and therefore infrequent. International expeditions were important in the 1960s and 1970s and are still used in the polar regions.

Ships move rather slowly (about 10 miles per hour) and can sample or observe only the ocean near them. In order to study ocean processes using such limited data, it was necessary to combine observations made decades apart and often separated by thousands of kilometers. The picture of the ocean that emerged from such sparse data was an ocean that changed little in time. If we had only such limited data for the land, we could determine the climate of dif-

FIGURE 1.8
The research vessel *New Horizon,* operated by the Scripps Institution of Oceanography.
Such ships are used for most oceanographic research and surveying. (Courtesy Scripps
Institution of Oceanography, University of California, San Diego)

ferent locations but could say little or nothing about weather changes over a
few hours. This is still one of the blindnesses we have about the ocean. We
know about its climate but are comparatively in the dark about its weather.
As we shall see, more frequent observations from satellites (Figure 1.9) are now
permitting more accurate descriptions and better predictions.

Studies of the ocean and its short-term interactions with the atmosphere
now use many different platforms and instruments. Studies of the deep-ocean
bottom and of its underlying structure require specialized drilling ships (Figure
1.10). Samples obtained by drilling deep-ocean sediments contain a record of
the Earth's climate and its changes over the past 200 million years. (Rocks
formed in more ancient oceans exist only on the continents.)

Studying ocean weather requires observation of the entire ocean within
a few days, which cannot be done using ships. Therefore, Earth-orbiting
satellites (Figure 1.9) now observe ocean processes. These platforms carry instruments to measure ocean-surface features such as temperature, roughness,
and color. Data from these instruments are relayed back to computers, which
combine the vast quantity of data into maps of the sea surface that are revised
every few days. Thus we now have maps for the ocean surface comparable

to our weather maps for the atmosphere. Fishermen receive daily reports on current locations and water temperatures.

One drawback to most instruments carried on satellites is that they can sense conditions only on the ocean surface. Other techniques are therefore necessary to study currents and other oceanic processes occurring in the deep ocean. (We discuss these techniques in Chapter 5.) Ships are now used as sampling laboratory platforms to supplement observations taken from space, and most studies are still done from them because of the satellites' inability to observe below the ocean surface. New techniques are being developed to sense the deep ocean from remote locations.

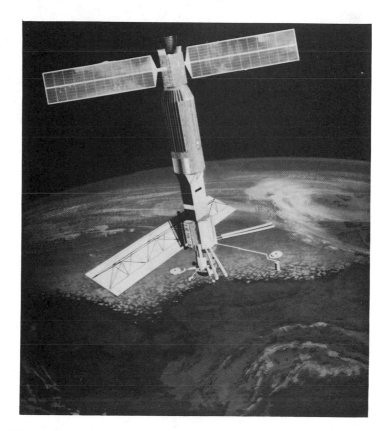

FIGURE 1.9
SEASAT-A was the first satellite to measure ocean conditions. The satellite circled the Earth 14 times a day, returning every 3 days to the same part of the ocean. Data from sensors on this satellite provided the first data on global ocean conditions. (Courtesy NASA)

FIGURE 1.10
Joides Resolution is equipped for scientific drilling to sample the deep-ocean bottom. The holes drilled are used for scientific studies and the cores are studied to learn about ocean history. (Courtesy Texas A & M University)

QUESTIONS

1. Explain the importance of the current around Antarctica.
2. List the important geographical feature(s) of each ocean basin. How does each feature influence the world ocean?
3. In which ocean basin are marginal seas most numerous?
4. Explain why the expansion or retreat of continental glaciers on land causes sea level to change.
5. Discuss the evidence for the antiquity of the ocean.
6. List and briefly discuss the processes that cause shoreline positions to change.
7. Which ocean basin receives the largest volume of river discharges? Which receives the least, relative to its volume?
8. Discuss the economic benefits to be realized from a detailed, accurate forecast of ocean currents.
9. What is the principal contribution of satellite remote sensing to the study of the ocean?

SUPPLEMENTARY READINGS

Books

Deacon, Margaret. *Scientists and the Sea: 1650–1900*. London: Academic Press, 1971. History of ocean sciences.

Dietrich, Gunter; Kalle, Kurt; Krauss, Wolfgang; and Siedler, Gerold. *General Oceanography: An Introduction*. 2d ed. New York: Interscience Publishers, 1980. General reference, good bibliography.

Linklater, E. *The Voyage of the Challenger*. London: John Murray, 1972. Account of the first modern ocean-exploring expedition.

Menard, H. W. *Anatomy of an Expedition*. New York: McGraw-Hill Book Company, 1969. Description of an oceanographic expedition.

Schlee, Susan. *The Edge of an Unfamiliar World: A History of Oceanography*. New York: Dutton, 1973. Emphasizes American oceanography.

Articles

Bailey, H. S., Jr. "The Voyage of the Challenger." *Scientific American* 188(5):88–94.

Barghorn, E. S. "The Oldest Fossils." *Scientific American* 244(5):30–54.

Browning, M. A. "Stick Charting." *Sea Frontiers* 19(1):34–44.

Bullard, Edward. "The Origin of the Oceans." *Scientific American* 221(3):66–75.

Kasting, James F.; Toon, Owen B.; and Pollack, James B. "How Climate Evolved on the Terrestrial Planets." *Scientific American* 258(2):90–98.

KEY TERMS AND CONCEPTS

Steady state assumption
Ocean basin boundaries
Continents
Distribution of land and ocean basins

Balance between precipitation, river discharge and evaporation
Changes in sea level

Effects of Ice Age
Ocean basin drilling
Computers
Satellites

2
Ocean Basins

The deep-ocean basins are still being explored. Their exploration began slowly in the nineteenth century, with sailing vessels lowering lead-weighted ropes or wires to determine their depth. Relatively few depth determinations, called soundings, were obtained in this way because of the difficulty and expense of making each measurement. With the limited data available, scientists thought that the ocean bottom was a flat, featureless plain. The shallow Mid-Atlantic Ridge was eventually discovered in the nineteenth century by surveys done for routing the transoceanic telegraph cables.

Our knowledge of the ocean bottom improved drastically after the **echo sounder**—developed in the 1920s to detect submarines—was used to map the deep-ocean floor. By determining the time between the sending of a sound pulse and the arrival back at the ship of its echo from the bottom, it was possible to map the ocean bottom more quickly and accurately. While the ship moved across an area, the echo sounder made a profile of the underlying bottom. By combining thousands of these profiles, it was possible to chart the ocean floor.

New techniques developed in the 1970s now map a swath of ocean bottom beneath the ship, using many echo sounders working simultaneously. Up to the early 1990s, still only about 20% of the bottom has been accurately mapped. Mapping newly claimed Exclusive Economic Zones along the ocean margins is one of the major tasks currently facing coastal nations (Figure 2.1).

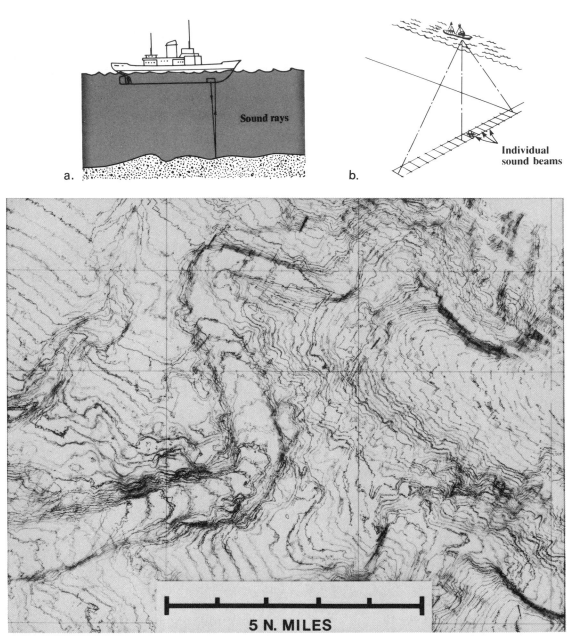

FIGURE 2.1

a. A single-beam echo sounder determines water depths directly beneath the ship. A sound pulse is reflected by the ocean bottom and the echo is recorded aboard ship. Water depth is determined by taking half the time elapsed between sending the pulse and receiving the echo and multiplying it by the speed of sound in water. b. A multi-beam echo sound uses many sound beams and a computer to process the echoes and to map a swath of ocean bottom beneath a survey ship. c. Bathymetric map prepared by swath-mapping techniques. (Courtesy National Ocean Survey)

2.1 OCEAN FLOOR TOPOGRAPHY

The ocean bottom is as irregular and as varied as the land. Its most conspicuous feature is the Earth-circling mountain range called the mid-ocean ridge. It also consists of enormous east-west tending cliffs, linear mountains (called fracture zones), and deep, narrow trenches near the ocean margins (Figure 2.2).

Nearest the continent is the **continental shelf**, averaging about 65 km (40 mi) wide. It slopes gently seaward and closely resembles the adjacent land. If coastal areas are rugged, so is the continental shelf, as, for example, off Southern California or Alaska. Where coastal regions have low hills or plains, adjacent continental shelves are similarly shallow and wide, as along the U.S. Atlantic or Gulf coasts (Figure 2.3).

The ocean deepens offshore over the **continental slope**. The **continental shelf break** marks the boundary between continental shelf and slope, where bottom slopes steepen, usually at depths around 130 m (430 ft). Both continental shelf and slope—referred to as the **continental margin**—are parts of continents. The exposed and submerged parts of the continents together account for about 44% of the Earth's surface.

Continental slopes are the edges of continents and form the largest continuous slopes on the Earth's surface. Pacific continental slopes are especially spectacular. Backed by some of the highest mountains on Earth, the total drop from the peaks of the South American Andes to the bottom of the Peru-Chile Trench is about 13,000 m (about 8 mi), nearly half again as high as Mount Everest (8850 m or 29,000 ft) in the Himalaya Mountains.

Many continental margins are cut by large **submarine canyons**. (Figure 2.4). Some are larger than Arizona's Grand Canyon of the Colorado River. Many are near mouths of large rivers such as the Congo, Hudson or Columbia; others are less obviously related. Some of these canyons are thought to have been cut by oceanic processes—dense, sediment-laden water masses called **turbidity currents** flowing along the seafloor.

Beyond the seaward edge of the continental slope is the **continental rise**, a large, featureless apron, sloping gently seaward. These aprons are built by sediment deposits derived from nearby continents. They form when sediment-transporting currents flow down submarine canyons and deposit their loads on the flatter seafloor, much like a mountain stream depositing sand and gravel as it emerges onto a flat valley floor. These aprons of sediment coalesce, forming thick deposits, known as the continental rise. Where sediment deposits are thick enough to bury any rough topography, they form smooth featureless bottoms called abyssal plains. Like the apron, abyssal plains slope gently away from the continents.

Deep ocean basins are typically 4000 to 6000 m (2.5 to 3.5 mi) deep, and they account for nearly 30% of the Earth's surface (Figure 2.5). Large areas of the deep-ocean bottom have an irregular floor, broadly arched with cones of

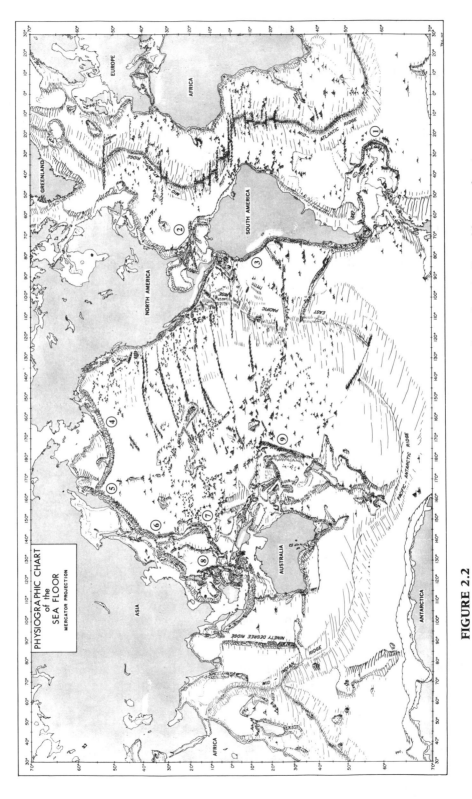

FIGURE 2.2
Topography of the ocean floor. Depths and names of trenches (indicated by numbers) appear in Table 2.1. (Copyright Hubbard Scientific Company. Used by permission)

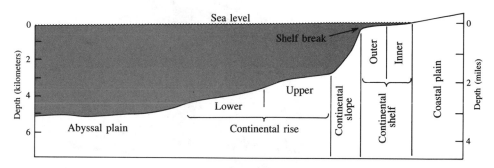

FIGURE 2.3
Schematic profiles of a continental margin, typical of the Atlantic coast.

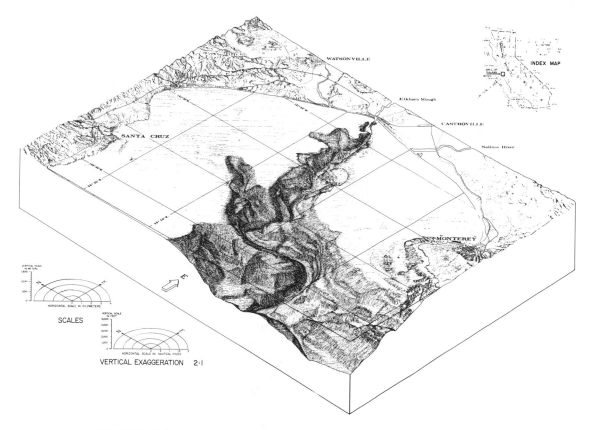

FIGURE 2.4
Monterey Canyon at the head of Monterey Bay, California, is one of the largest submarine canyons on the U.S. Pacific coast. (Courtesy Tao Rho Alpha, U.S. Geological Survey)

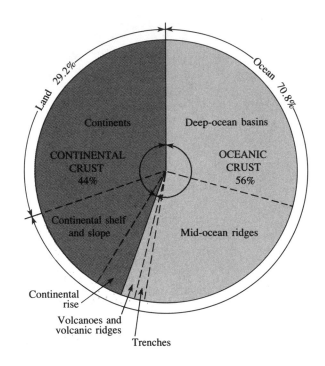

FIGURE 2.5
Fraction of the Earth's surface composed of continental and oceanic crust.

submarine volcanoes. They are also marked by steep-sided valleys bounded by faults—fractures in rocks where one side has moved relative to the other.

Mid-ocean ridges constitute nearly 23% of the Earth's surface (Figure 2.5) and occur in all major oceans. Some lie near the center of the ocean basin, for example, the rugged **Mid-Atlantic Ridge** and the **Mid-Indian Ridge** (Figure 2.2). But in the Pacific, the **East Pacific Rise** lies near the South American continent. It intersects North America in the Gulf of California and reappears near the California-Oregon boundary. Some volcanic islands (Iceland, for example) are exposed parts of mid-ocean ridges.

There are many low hills or **knolls** rising less than 1000 m above the seafloor where sediment deposits are too thin to bury irregular topography. These hills are small, extinct volcanoes on the seafloor. Large volcanoes also occur, individually and in groups, especially in the western Pacific. The tops of the largest volcanoes form islands, such as the **Hawaiian Islands**. Rising nearly 9 km (about 5.6 mi) from the seafloor, the volcanoes forming the Hawaiian chain are the largest mountains on Earth.

Trenches occur around most of the Pacific Ocean basin. A few occur in the Atlantic and northern Indian oceans (Figure 2.2). These long, narrow depressions are the deepest parts of the ocean floor. Several western Pacific trenches

TABLE 2.1
Maximum Depths of Trenches

	Depths	
	(Meters)	(Feet)
ATLANTIC OCEAN		
South Sandwich trench (1)*	8,400	27,600
Puerto Rico trench (2)	9,200	30,000
PACIFIC OCEAN		
Peru-Chile trench (3)	8,050	26,400
Aleutian trench (4)	8,100	26,500
Kuril-Kamchatka trench (5)	10,500	34,400
Japan trench (6)	9,800	32,100
Mariana trench (7)	11,000	36,000
Philippine trench (8)	10,000	33,000
Kermadec-Tonga trench (9)	10,800	35,400
INDIAN OCEAN		
Java trench (10)	7,460	24,400

Source: U.S. Naval Oceanographic Office; and R. L. Fisher and H. H. Hess, ''Trenches,'' *The Sea: Ideas and Observations on Progress in the Study of the Seas,* vol. 3 (New York: Interscience Publishers, 1963), pp. 418–19.
*Trench locations are indicated by numbers on Figure 2.2.

are nearly 11 km (about 36,000 ft) deep (Table 2.1), and broad areas of low relief roughly parallel their seaward sides.

Arcuate groups of volcanic islands, called **island arcs,** are associated with most trenches. These areas are subject to large, destructive earthquakes, and many of their volcanoes are active. Island arcs are most common in the western Pacific, although one occurs in the south-western Atlantic Ocean and another in the northern Indian Ocean. Shallow marginal seas occur between continents and some of the largest island arcs, such as the Japanese and Philippine islands.

Fracture zones are long belts of steep cliffs, rugged topography, and sea-mounts, which are formed by large faults that cut the ocean floor. They extend for thousands of kilometers across all the ocean basins and are especially conspicuous in the Pacific.

2.2   PLATE TECTONICS

The Earth's crust is broken into several large, rigid units called **lithospheric plates.** *Each plate moves as a unit in a process called* **plate tectonics** *or* **seafloor spreading.** In this process, the cold, rigid lithosphere moves across a nearly molten layer called the **asthenosphere.** New lithosphere forms at mid-ocean ridges. The oldest lithosphere is destroyed at trenches. Plate movements are

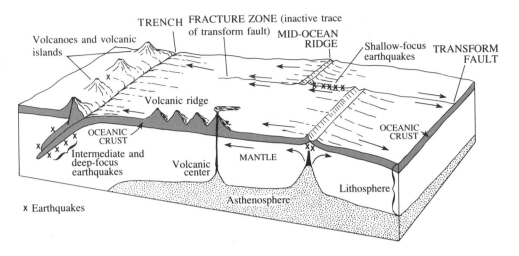

FIGURE 2.6
Schematic representation of lithospheric plate movements. New oceanic crust is formed by mid-ocean ridges and destroyed at trenches.

slow, typically 2 to 3 cm (about 1 in) per year. (This is about the same rate as your fingernails grow.) In the South Pacific, the movements are much faster, annually moving up to 20 cm (8 in). Regardless of the speed, the forces involved are strong enough to break up continents and regroup the pieces (Figure 2.6).

There are seven large plates—each larger than individual continents—and many smaller plates (Figure 2.7). Most of them include parts of continents as well as ocean basins. *Their boundaries fall into three types: (1) mid-ocean ridges where plates form, (2) trenches where plates are destroyed, and (3) transform faults where plates slide past each other.*

Plate tectonics helps us explain many enigmatic facts, among them the close fit between the opposite sides of the Atlantic Ocean. About 200 million years ago, a large continental mass—called **Pangaea**—broke up. Through plate tectonics the pieces moved irregularly to their present locations.

Plate movements are caused by motions of the underlying mantle. The plates are rigid so there is little faulting (cracking of the crust), volcanic activity, or mountain building within each one. Edges of plates are marked by earthquakes and active volcanoes. In fact, earthquakes are used to map plate boundaries.

2.3   MID-OCEAN RIDGES

New lithosphere forms at mid-ocean ridges by volcanic eruptions and outflows of molten rock through fissures. The fissures open up as the plates move away from the ridge axis. Shallow earthquakes occur at ridge crests, probably associated with volcanic eruptions, and volcanic islands form on the Mid-Atlantic and Mid-Indian ridges. Iceland, with its active volcanoes, is an exposed part of the Mid-Atlantic Ridge.

Oceanic crust forms by intermittent volcanic eruptions in the central valley of the mid-oceanic ridges. As hot lava (about 1200°C) contacts cold seawater (2–5°C), its outer surface congeals. The resulting lava flows assume characteristic tube- like and pillow-shaped forms (Figure 2.8). Larger volcanic eruptions form sheet flows that have chilled upper surfaces under which liquid lava flows.

Where crustal formation is relatively slow (low spreading rates are typically 6 cm/yr) the mid-ocean ridge has a distinct **rift valley** about 1 km wide. More rapid spreading results in a broad dome-like rise with no rift valley. Thus the slow spreading Mid-Atlantic Ridge has a rift valley and relatively rugged topography, and the rapidly spreading East Pacific Rise generally has a broad dome, sometimes with a central bulge (Figure 2.2) but rarely a central rift valley.

Heat is removed from newly formed crust by two processes. **Conduction** through the ocean floor removes about one-third of the heat lost through the mid-ocean ridge. The rest is removed by seawater circulating through fractured rocks. Warm waters flow out through cracks and fissures in the ocean bottom, while hot waters circulate many kilometers below the sea bottom. This process is called **hydrothermal circulation**.

Where spreading rates are relatively rapid (more than 6 cm/yr), much of the hot water (temperatures up to 400°C) discharges through vents. Minerals precipitate out of the hot waters as they mix with cold, oxygen-containing ocean waters and form conical structures up to 10 m high (Figure 2.9). These deposits of copper and zinc are mined where they occur on land, such as on the island of Cyprus in the Mediterranean.

2.4 AGE AND DEPTH OF OCEAN FLOOR

Magnetic minerals in rocks record magnetic pole reversals and plate movements. As lava cools, newly formed magnetic minerals record the Earth's magnetic field at that time. The orientation of these minerals in the Earth's crust can be determined by towing sensitive instruments called **magnetometers** behind ships or planes. Minerals can also be studied in rocks taken from the ocean bottom.

Earth's magnetic field reverses itself at irregular intervals. When this happens, the north and south magnetic poles change places, and magnetic minerals in rocks forming at mid-ocean ridges record the new magnetic direction. Since oceanic crust is continually formed at mid-ocean ridges, the ocean bottom continually records the orientation of the magnetic field. Thus *the rocks forming the oceanic floor act as a gigantic tape recorder, preserving a record of changes in the Earth's magnetic field.*

Volcanic rocks formed on land also record changes in the Earth's magnetic field, and a time-scale of its reversals has been established. This has allowed oceanographers to use the pattern of magnetic reversals to interpret the magnetic

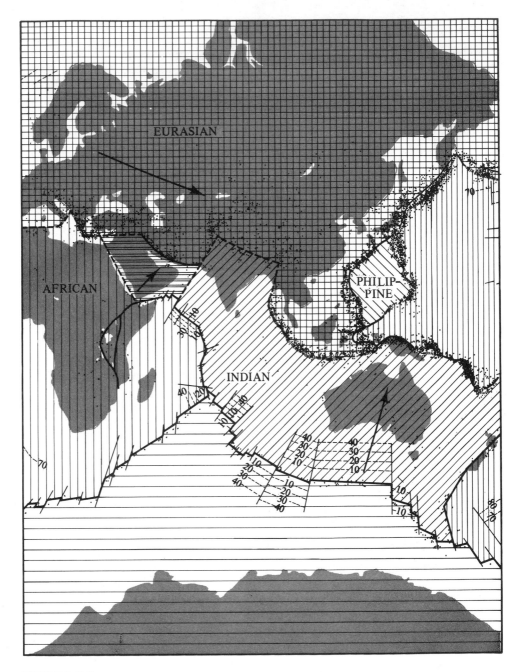

FIGURE 2.7

Earth's lithosphere consists of seven major and many smaller plates, each moving as a rigid unit. Boundaries between plates are mid-ocean ridges (or rises) where crust forms; trenches (or folded mountain belts) toward which they converge; and transform faults, where plates slide past each other. Volcanic centers form volcanic ridges and island chains where plates move across a volcanic center. Arrows indicate directions of plate movements; numbers indicate speed in centimeters per year.

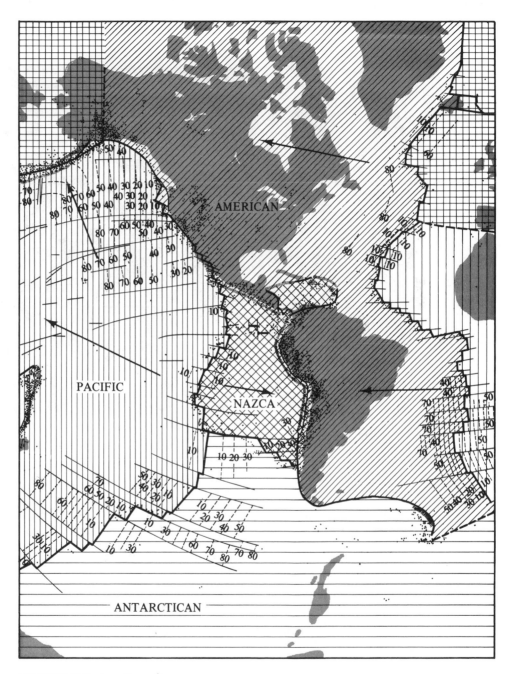

FIGURE 2.7 (*continued*)

FIGURE 2.8
Lava from seafloor volcanic eruptions form characteristic tubelike and pillow-shaped forms on the ocean floor, as photographed from the submersible ALVIN. (Courtesy Woods Hole Oceanographic Institution)

pattern observed in oceanic crust to determine when the ocean bottom formed. From these magnetic patterns they reconstruct how ocean basins formed and how continents have moved.

These magnetic patterns show that much of the Atlantic Ocean bottom took shape in the past 80 million years: Its oldest part is about 160 million years old (Figure 2.10). These magnetic patterns also provide details on ocean basin formation. For example, we know by them that the North Atlantic formed before the South Atlantic. The oldest rocks in the North Atlantic are 160 million years old; the oldest in the South Atlantic are only 140 million years old (Figure 2.10).

The newly formed ocean floor at mid-ocean ridges stands higher than its surroundings because the crust is hotter and less dense. As the plate moves away from the ridge, the crust cools, becomes denser, and sinks. Thus the older the crust, the deeper the ocean floor. In other words, *the ocean floor acts much like a conveyor belt that gradually deepens as it ages*. At mid-ocean ridges, the ocean bottom is about 2500 m deep. It is about 6000 m deep when it is 160

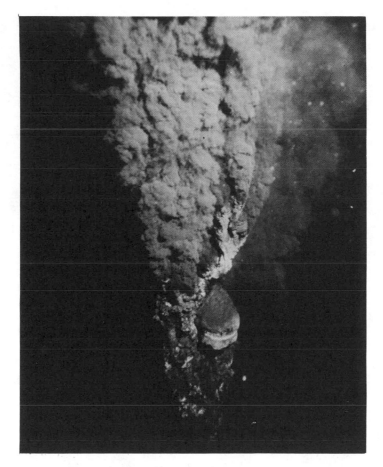

FIGURE 2.9
Hot waters (350°C) discharged from a "black smoker" on the East Pacific Rise. Sulfides in vent waters react with seawater, precipitating metal-sulfide particles, which accumulate in nearby sediment deposits. (Photograph by Robert D. Ballard. Courtesy of Woods Hole Oceanographic Institution)

million years old (Figure 2.11). The oldest oceanic crust is in the western North Pacific. It is about 190 million years old. Older crust may occur in some of the small marginal ocean basins but none has yet been sampled.

2.5   TRENCHES AND ISLAND ARCS

Lithosphere is destroyed at trenches (Figure 2.6 and 2.12a) *where oceanic plates are pulled down beneath adjacent plates.* Both the plate and the sediment deposited on it are drawn into the mantle and eventually melted and mixed

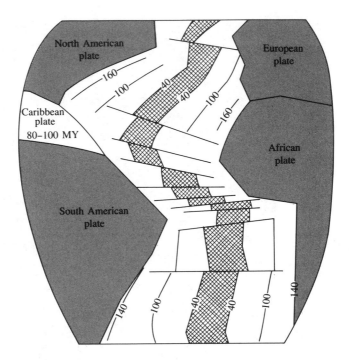

FIGURE 2.10
Age of oceanic crust of the Atlantic Ocean, based on patterns of magnetic field reversals in ocean-floor rocks.

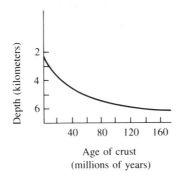

FIGURE 2.11
The depth of the ocean floor increases with age as the crust cools. [After J.G. Sclater, R.N. Anderson, and M.L. Bell. The elevation of ridges and evolution of eastern Pacific. *Journal of Geophysical Research* 76 (1971), pp. 7888–7915.]

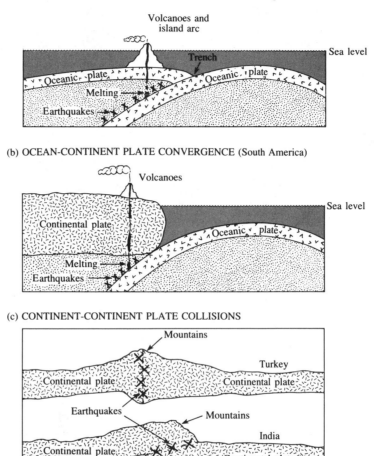

(a) OCEAN-OCEAN PLATE CONVERGENCE (Aleutians, Japan)

Volcanoes and island arc

Trench

Sea level

Oceanic plate   Oceanic plate

Melting

Earthquakes

(b) OCEAN-CONTINENT PLATE CONVERGENCE (South America)

Volcanoes

Sea level

Continental plate

Oceanic plate

Melting

Earthquakes

(c) CONTINENT-CONTINENT PLATE COLLISIONS

Mountains

Turkey

Continental plate   Continental plate

Earthquakes   Mountains

India

Continental plate   Continental plate

FIGURE 2.12
Schematic representation of different types of plate convergences.

with other mantle materials in a process called **subduction**. Melting of lithosphere and sediment in the mantle causes volcanism above the subduction zone.

Where an oceanic plate converges with another oceanic plate, the abundant volcanic activity of subduction zones forms island arcs. Usually the trench lies on the seaward side of the chain with a relatively shallow sea nearest the continent. These shallow marginal seas often accumulate thick sediment deposits. High heat-flows from the young crust under marginal seas provide especially favorable conditions for forming oil and gas deposits.

Trenches and island arcs are common around the Pacific, especially in the western Pacific. The only gaps in the circum-Pacific trench-island-arc systems are the mountains of western North America and Antarctica.

Deep earthquakes occurring at depths greater than 100 km are characteristic of trenches. Movements of the lithosphere that cause earthquakes occur on planes dipping beneath adjacent crustal blocks (Figures 2.6 and 2.12). The deepest earthquakes occur at depths of 700 km. On transform faults and on mid-ocean ridges, earthquakes occur at relatively shallow depths, from a few kilometers to a few tens of kilometers deep.

Continents can also converge, but these convergences do not form trenches or island arcs. The most conspicuous convergence is the Mediterranean, where Africa and Europe are now colliding. South of Greece there is only a small trench, the **Hellenic Trench,** but no obvious island arc. In Turkey and Pakistan, the convergence is marked by destructive earthquakes and mountain building, and within a few million years, Australia will collide with southeast Asia.

In the last 40 million years, the Indian subcontinent has also moved thousands of kilometers northward and collided with Asia. The most conspicuous result is the Himalaya Mountains, relatively young mountains formed where part of the Indian block has been thrust beneath the Asian block. The Rocky Mountains of Western North America were formed by subduction of the Pacific plate beneath North America. The height of these mountains results from the buoyant, relatively young oceanic plate lying beneath the continental crust.

2.6   TRANSFORM FAULTS

Locations where lithospheric plates slip past each other along transform faults are marked by fracture zones. A line of earthquake epicenters or areas of volcanic activity marks the active part of transform faults (Figure 2.6).

Earthquakes occur on transform faults only between ridge crests. In this segment of the fault there are differential movements of plates across the fault. Beyond ridge crest segments, the plates on each side of the transform fault are moving in the same direction and at the same speed so there are no earthquakes. Traces of transform faults persist for many millions of years, until buried by sediment deposits. Over time they record ancient movements of plate segments. These fracture zones are especially conspicuous in the Pacific where they remain unburied by sediments.

2.7   CONTINENTAL MARGINS

Continental margins mark transitions between granitic, continental crust and the denser, basaltic, oceanic crust. They are formed by the same plate movements that shape the ocean basins. There are three major types of continental margins, distributed as shown in Figure 2.13.

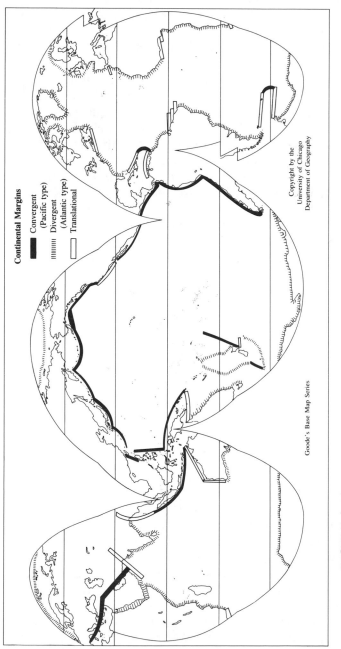

FIGURE 2.13

Distribution of different types of continental margins. [After K.O. Emery. Continental margins—Classification and Petroleum Prospects. *Bulletin of the American Association of Petroleum Geologists* 64 (1980), pp. 297–315]

Divergent margins (also called passive or Atlantic-type) form when continents are pulled apart, forming new ocean basins. These continental margins mark the transitions between oceanic and continental crust, but they are not plate boundaries. The passive continental margins around the Atlantic are conspicuous examples.

Divergent margins form by the thinning or faulting of continental crust at newly formed mid-ocean ridges. Through time the continental fragments move apart and gradually subside as they cool. Because of their subsidence, they accumulate thick sediment deposits. Continental slopes and rises contain nearly two-thirds of all sediment deposits in the ocean. These sediment accumulations are especially favorable locations for forming oil and gas deposits.

Convergent margins *(also called* **active** *or* **Pacific-type***) mark the boundaries between converging crustal plates.* This is sometimes associated with subduction and can result in an oceanic plate thrusting under a continental plate, as is now occurring along the Pacific coast of North and South America, or in one continental block underthrusting another, as in the Himalaya Mountains of northern India. Marginal ocean basins occur along the convergent plate boundary of the Asian Pacific coast.

Translational margins (also called transform margins) *are formed by lateral motions between plates.* They may be marked by shallow earthquakes. During rifting, parts of continental crust move relative to the adjacent crustal plate. At present, parts of California (west of the San Andreas Fault) and Baja California are moving northwest relative to the North American plate. Eventually parts of Southern California and Baja California will form a large island offshore from the rest of North America. The Channel Islands now offshore of Southern California are pieces broken off the continent many million years ago.

2.8   VOLCANOES AND VOLCANIC ISLANDS

Many mountains on the ocean bottom are extinct volcanoes, as are many islands. Some ancient, extinct volcanoes are now submerged and covered by coral reefs and deposits of carbonate sediments.

Volcanoes form when molten rock erupts at the Earth's surface. Through time, lava and ash from these eruptions accumulate, forming a mountain. About 20 major volcanic eruptions occur each year on land. A much larger number of eruptions occurs on the ocean bottom, usually unobserved. Only shallow submarine eruptions (Figure 2.14) and those occurring on islands have been recorded and studied.

Unless renewed by continued eruptions, volcanic islands are eroded down to sea level by weathering and wave action. Intermittent volcanic activity often forms disappearing islands. These are volcanic cones built on submerged banks. They are worn down by waves or destroyed by explosions associated with later eruptions. As we discussed previously, the volcano subsides as the plate on which it rests subsides through time.

FIGURE 2.14
The island of Surtsey as it appeared in the early 1970s. Unless volcanic eruptions continue, the island will be eroded away by waves and rain until only a shallowly submerged bank remains. (Photograph courtesy Icelandic Airlines)

Most of the Earth's most active volcanoes occur on plate boundaries, such as the volcanoes of the island arcs in Japan or Indonesia. Some of the largest volcanoes occur on or near mid-ocean ridges, such as Iceland, the Azores, or Ascension Island in the South Atlantic.

Long-lived volcanic centers, called **hot spots,** cause lines of volcanoes in the interior of plates as the plates move across them. Active volcanoes occur immediately above the hot spot, and as the plate continues to move, the volcanic cone is moved away from the hot spot and eventually becomes extinct. A new cone forms above the hot spot and continues to be active until it too moves too far from the source of molten rock. In this way a chain of volcanic islands or a volcanic ridge forms on the seafloor.

The most conspicuous example of islands formed by a plate moving over a hot spot is in Hawaii (Figure 2.15). The active Hawaiian volcanoes are at the southeastern end of the chain; the oldest and most eroded islands, now marked by coral atolls, lie to the northwest. The long line of islands marks the plate's movement across the hot spot. A new, still submerged, volcanic cone, called Loihi, is now forming southeast of Hawaii. A chain of deeply submerged seamounts, the **Emperor Seamounts,** extends north-south on the Pacific Ocean bottom. This marks the trace of the plate over the same hot spot before the Pacific plate changed direction about 40 million years ago.

Growth of a volcano, whether on land or on the ocean bottom, is not a continuous process. Large explosions, some caused by water coming in contact with molten rock, can blast away parts of the mountain. In 1883, Krakatoa, a volcano between Java and Sumatra in Indonesia, caused one of the largest

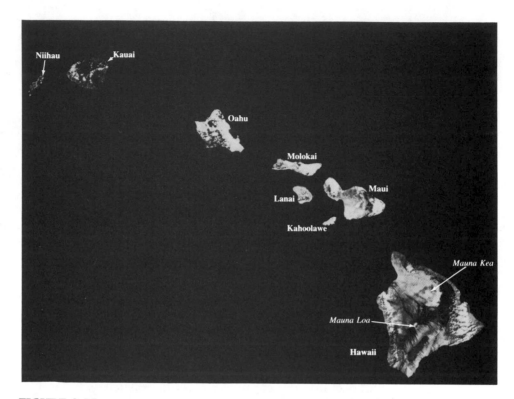

FIGURE 2.15
The youngest and largest of the Hawaiian Islands occur at the eastern end of the chain.
The island of Hawaii has two active volcanoes, Mauna Loa and Mauna Kea. Kauai,
the oldest island shown here, has been most eroded by waves and weathering. (Courtesy
NASA)

volcanic explosions in historic time. Following 200 years of inactivity, the
volcano first had several weeks of moderate activity. Then on August 26, a
violent explosion, heard 5000 km away and measured around the world, began
two days of explosions that obliterated two-thirds of the island. The explosions
caused gigantic ocean waves, called **tsunamis** (which we discuss further in
Chapter 6), which destroyed nearby low-lying islands, killing 36,000 people.
The waves reached as high as 40 m (130 ft) above sea level and up to 16 km
(10 mi) inland on Java. Tsunamis wasn't the only consequence, however. Wind-
blown volcanic ash formed thick deposits over 770,000 km² (300,000 mi²), and
thinner deposits covered 3.8 million km² (1.5 million mi²). About 16 km³ (3.8
mi³) of ash was blown into the air, and deposits were 15 m (50 ft) thick on
nearby islands. Large amounts of ash were doubtlessly deposited on the deep-
ocean floor, as well as being carried 50 km (30 mi) into the stratosphere, where
it remained for several years. During that time it caused brilliantly colored
sunsets throughout the Northern Hemisphere.

A similar huge volcanic explosion on the Greek island of Santorini around 1500 B.C. may have given rise to the Atlantis legend. That ancient "lost city" may have been destroyed by an enormous explosion and tsunamis. The widespread destruction may have hastened the decline of the Minoan civilization and the subsequent rise of classical Greek culture.

2.9   SUPERCONTINENT CYCLES

Movements of lithospheric plates are driven by heat escaping from the Earth's interior. This heat comes from the decay of radioactive elements in the Earth's core and mantle and from the original formation of the Earth.

Earth has been cooling since its formation. Since the crust formed about 3,800 million years ago, Earth has undergone six, or maybe seven, cycles of supercontinent formation. Supercontinents break up, forming a new ocean which grows and later disappears, forming another supercontinent. The cycle from start to finish takes about 500 million years (Figure 2.16).

Continental crust does not conduct heat as well as oceanic crust. Thus a continent remaining in one location for a long time leads to heating of the underlying mantle because it retards heat flow. In short, continental crust acts like a blanket.

As the mantle warms, it expands, elevating the overlying continent and stretching its crust. Eventually the crust breaks, forming a rift valley. An example of this process can be seen in Africa, which has been in its present location for about 200 million years and stands about 400 m higher than other continents. As its rift valley grows, it forms a long narrow ocean. (An example is the Red Sea, which is becoming a narrow ocean basin.)

Eventually a mid-ocean ridge forms. This causes the narrow basin to widen. Through time, the oldest part of the lithosphere in the ocean basin becomes dense enough to sink into the mantle. This is now happening in the South Atlantic and in the West Indies. Basins typically widen for about 200 million years and close after the mid-ocean ridge is subducted.

As the ocean basin ages, subduction becomes more widespread. Eventually the basin begins to close as subduction overwhelms spreading due to the mid-ocean ridge. For about 200 million years, the ocean basin closes. As the Atlantic-type ocean closes, sediment deposits in the basin are deformed, resulting in a mountain range on the newly assembled supercontinent that marks the site of the former ocean. The Appalachian Mountains were formed this way during the previous cycle about 450 million years ago.

The Pacific Ocean remains intact during these cycles. As the Atlantic-type ocean widens, subduction occurs to accommodate the lithospheric blocks that are moving over its margins. It is now bordered by subduction zones in all but a few locations. When the Atlantic-type basin begins closing, most subduction in the Pacific will cease.

The supercontinent cycle also affects sea level relative to the continents.

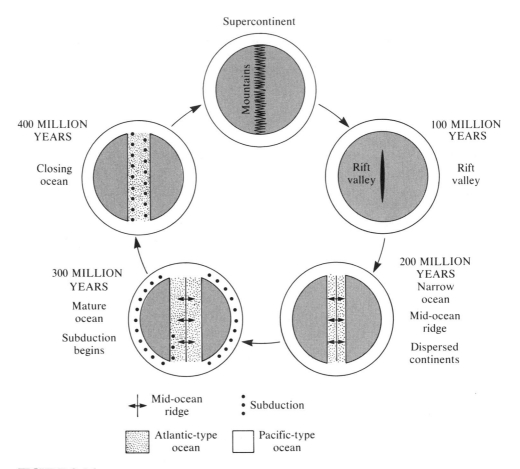

FIGURE 2.16
Supercontinents form and break up to form new ocean basins, which then close to reform supercontinents.

During the supercontinent phase, the continents stand high; as spreading begins, the continents move off the heated mantle and thus stand lower. At this time, there is extensive flooding of low-lying areas, and the newly formed ocean basin floor is relatively shallow. As the ocean basin widens and ages, sea level falls.

QUESTIONS

1. What fraction of the Earth's surface is underlain by continental materials?
2. What are the three types of plate boundaries? Briefly describe the principal features of each.
3. List the major parts of the continental margin.
4. Describe island arcs and their origin.
5. Which type of plate boundary is characterized by deep-focus earthquakes?

6. Briefly describe and compare the structures and composition of ocean basins and continental blocks.
7. Which seafloor features are part of the continents? Which are parts of the ocean basins? How are they distinguished from each other?
8. Briefly describe the basic process involved in plate tectonics.
9. In the western Pacific Ocean, what seafloor features commonly mark the boundary between ocean basin and continent?
10. What causes volcanic activity at subduction zones?
11. Discuss the relationships between crustal age and ocean depths.

SUPPLEMENTARY READINGS

Books

Kennett, James P. *Marine Geology*. Englewood Cliffs, N.J.: Prentice-Hall, Inc., 1982. Comprehensive, technical; assumes geological background.

Miller, Russell. *Continents in Collision*. Alexandria, VA: Time-Life Books, 1983. Well illustrated elementary treatment of plate tectonics, including development of the theory and history of plate movements.

Sullivan, Walter. *Continents in Motion: The New Earth Debate*. New York: McGraw-Hill Book Company, 1974. Excellent discussion of the origins of plate tectonic theory.

Wyllie, Peter J. *The Way the Earth Works*. New York: John Wiley & Sons, Inc., 1976. Elementary, comprehensive.

Articles

Dewey, J. F. "Plate Tectonics." *Scientific American* 266(5):56–68.

Emery, K. O. "Continental Shelves." *Scientific American* 221(3):106–125.

Heezen, B. C. "Origin of Submarine Canyons." *Scientific American* 229(5):102–112.

Heezen, B. C., and MacGregor, I. D. "The Evolution of the Pacific." *Scientific American* 251(1):46–55.

Hekinian, Roger. "Undersea Volcanoes." *Scientific American* 251(1):46–55.

Hoffman, K. A. "Ancient Magnetic Reversals." *Scientific American* 258(5):76–83.

Matthews, S. W. "This Changing Earth." *National Geographic* 143:1–37.

Menard, H. W. "The Deep-Ocean Floor." *Scientific American* 221(3):126–145.

Nance, R. D.; Worsley, T. R.; and Moody, J. B. "The Supercontinent Cycle." *Scientific American* 259(1):72–77.

Sclater, J. G., and Tapscott, C. "The History of the Atlantic." *Scientific American* 240(6):156–175.

KEY TERMS AND CONCEPTS

Plate tectonics	Mid-ocean ridges	Volcanic centers (hot spots)
Ocean bottom topography	Hydrothermal vents	Volcanoes
Submarine canyon	Trenches	Tsunamis
Plate divergences	Transform faults	
	Continental margins	

3
Seawater

Water is the most abundant substance on the Earth's surface, and its properties have a profound effect on the ocean's chemical, physical, and biological makeup. To understand the behavior of the ocean we need to understand seawater. In this chapter, we start with pure water and its properties and then examine the effects of sea salt. In later chapters we will consider the effects of life on the ocean.

3.1 WATER MOLECULES

Each water molecule consists of an oxygen atom bonded to two hydrogen atoms. Because of the oxygen atom's electron cloud configuration, hydrogen atoms in water molecules occur at two corners of a four-cornered (tetrahedral) molecule (Figure 3.1). A water molecule has two hydrogen atoms on one side, and each hydrogen atom shares its single electron with the oxygen atom. Hydrogen atoms thus act as a positive charge on the water molecule. On the opposite side is the oxygen atom, which, because of the shared electrons, has excess negative charges. *Since water's positive and negative charges are separated, it is called a* **polar molecule**.

All molecules attract each other. The attractions are weak and arise from interactions between the atomic nuclei of one molecule and the electrons of another when the two molecules are close together. In water, the separation of charges and the presence of hydrogen atoms together give rise to stronger

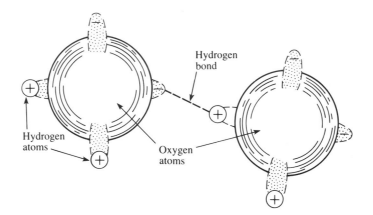

FIGURE 3.1
Water molecules have a four-cornered structure. The molecules are shown linked by a hydrogen bond, an interaction between a hydrogen atom of one molecule and a negative charge on the other water molecule.

interactions known as **hydrogen bonding**. Hydrogen bonds between water molecules are only about 1/20 as strong as the bonds between hydrogen and oxygen within the molecule itself. Nonetheless, they are strong enough to influence the properties of liquid water. If hydrogen bonding were not present, ice would melt at about −100°C (−148°F) and water would boil at about −80°C (−112°F). In other words, water would exist only as a gas at Earth's surface temperatures and pressures. *Without hydrogen bonding of water molecules, there would be no ocean and no life on Earth.*

3.2   FORMS OF WATER

Water is one of the few common substances on Earth which exists in all three **forms of matter: crystalline solid**—ice; **liquid**—water; and **gas**—water vapor. We will consider each form to see how its structure controls its properties.

Like all crystalline solids, ice has an orderly internal structure (Figure 3.2) in which each molecule is bound so it can neither move nor rotate freely. Its intermolecular bonds are somewhat elastic (like springs), permitting molecular vibrations but inhibiting long-range movement.

Because of its structure, ice is a rather open network of water molecules. The molecules in ice are not as closely packed as a similar number of molecules in liquid water, which fit together like marbles in a cup. Ice at 0°C has a density of about 0.92 g/cm³; liquid water at the same temperature has a density of about 1 g/cm³. The different densities explain why any volume of ordinary ice will float in an equal volume of water.

Sea ice contains less salt than does seawater. Despite the openness of the ice structure, most impurities will not fit between its molecules. Since salts do

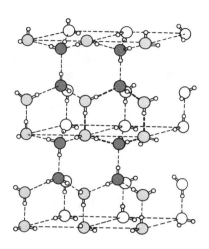

FIGURE 3.2
Crystal structure of ice. Note the six-sided rings formed by the water molecules.

not substitute readily for water molecules in ice, sea salts and dissolved gases are excluded from ice when sea water freezes.

When ice melts, liquid water forms. From observing the disappearance of ice crystals during melting, we might expect that all intermolecular bonds responsible for the ice structure would also disappear. But this is not the case. Instead, many water molecules remain bound together in clusters (Figure 3.3) surrounded by unbonded water molecules. As a result, water is a highly atypical liquid. In fact, it behaves like a crystalline substance at the temperatures found in the ocean. This results from the strong tendency of water molecules to form

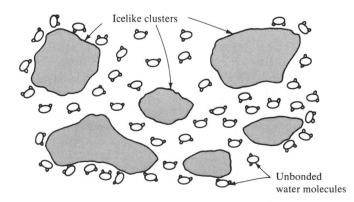

FIGURE 3.3
Schematic representation of the liquid water structure. [After R.A. Horne, "The physical chemistry and structure of sea water." *Water Resources Research* 1 (1965), p. 269]

hydrogen bonds. At low temperatures, many of the molecules in liquid water occur in clusters at any instant (Figure 3.3). The proportion of bonded molecules and the relative size of the cluster decrease as temperatures rise (Figure 3.4). At the boiling point, all bonding between molecules breaks down.

Liquid water, unlike ice, flows readily, maintaining only a fixed volume at a given temperature. It can flow because some of its water molecules can move and rotate freely. These unbound water molecules form one or two layers of closely packed molecules (Figure 3.3) that can move or rotate with little or no restriction around the cluster.

Liquid water's structure is far from static. Its molecules alternate rapidly between structured and unstructured states. The intermolecular bonds that cause its clusters to hold together break up and re-form about 10^{12} (one million million) times each second. In ice, bonds persist much longer, breaking about once each second.

The relative abundance of structured and unstructured components of water change with varying temperatures, pressures, and salt contents. Rapid changes or flicker of the structure account for water's ability to flow. If the structure did not rapidly break and re-form, water would be as rigid or brittle as ice, fracturing instead of flowing.

Water vapor, a gas, has neither shape nor size but completely fills any container in which it is placed. For liquid water to change to water vapor, all intermolecular bonds must be broken. Individual molecules may then move and rotate independently.

On a molecular level, water vapor in a container may be compared to a room full of angry bees. Each molecule moves independently, unaffected by

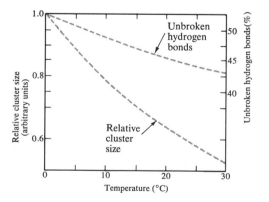

FIGURE 3.4
Effect of temperature on the relative number of unbroken hydrogen bonds and cluster size in pure water. [Data from G. Nemethy and H.A. Scheraga, "Structure of water and Hydrophobic Bonding in Protein." *Journal of Chemical Physics* 36 (1962), p. 3394]

the others. The pressure that a gas exerts on its container results from molecules colliding with the walls. As temperature rises, the molecules move more rapidly, and their collisions with the walls increase; in other words, the **pressure** increases with higher temperatures. In a vapor there is much space between individual molecules. Thus, we can add other gases with little difficulty, although total gas pressure increases.

3.3 SALTS IN SEAWATER

Seawater is a solution of salts of nearly constant composition, dissolved in variable amounts of water. Water, its most abundant constituent, determines most of seawater's physical properties. Salts in seawater control its density and other of its physical characteristics.

Despite the large number of elements (more than 70) dissolved in seawater, *only six of them (Figure 3.5) make up more than 99% of all sea salts*: chlorine (Cl); sodium (Na); magnesium (Mg); calcium (Ca); potassium (K); and sulfur (S). Common table salt (NaCl) alone makes up nearly 86%. All occur as **ions**—electrically charged atoms or groups of atoms. When substances dissolve in water, electrons are stripped from some constituents, which then form positive ions. The constituents that retain their electrons have an excess amount, and therefore are negatively charged. The number of positive and negative ions in a solution must balance.

Oceanographers use **salinity**—*the amount (in grams) of total dissolved salts present in one kilogram of water*—to express the salt content of seawater. Salinity (S) is determined by measuring the electrical conductivity of a seawater sample: the higher the conductivity, the greater the salinity.

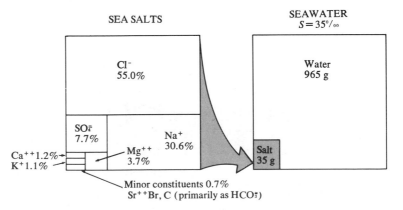

FIGURE 3.5
Relative proportions of water and dissolved salts in seawater.

3.4   DISSOLVED GASES

Seawaters contain small amounts of dissolved gases (Figure 3.6). Because of the constant stirring of the sea surface by winds and waves, atmospheric gases dissolve in surface waters. *Water of a given temperature and salinity is saturated with gas when the amount of gas entering the water equals the amount leaving during the same time. Surface seawater is normally saturated with atmospheric gases such as oxygen and nitrogen.*

The amount of gas that can dissolve in seawater is determined by the water's temperature and salinity. Increasing the temperature or salinity reduces the amount of gas that can be dissolved. Of the two components, temperature is the more important. Like temperature and salinity, the gas dissolved in a bit of seawater is controlled by conditions existing where the water was last at the surface.

Once water sinks beneath the ocean surface, dissolved gases can no longer exchange with the atmosphere. Two things may then happen. The amount of gas in a bit of water may remain unchanged, except by movement (diffusion) of gas molecules through the water—a slow process—or by the water's mixing with other water masses containing different amounts of dissolved gas. In general, nitrogen (N_2) and the rare, chemically inert gases in the atmosphere behave this way. We say that their concentrations are **conservative properties,** affected only by physical processes. (Salinity is another example of a con-

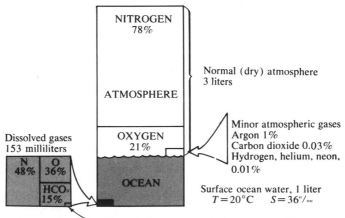

FIGURE 3.6
Gases dissolved in saturated seawater in contact with a dry atmosphere. For each liter of seawater, there are about 3 liters of atmosphere.

servative property of seawater.) Seawater is saturated with nitrogen and rare gases since they are unaffected by chemical or biological processes.

In addition to mixing and diffusion, some gases participate in biological or chemical processes that change their concentrations. They are examples of **nonconservative properties**. For instance, oxygen and carbon dioxide are released and used at varying rates in the ocean. Variations in the concentrations of such gases can be used to trace subsurface water movements. Compared to other atmospheric gases, the amount of carbon dioxide is unusually large because of chemical reactions in water involving carbon dioxide. Chemical reactions involving carbon dioxide play a major role in controlling the acidity of seawater.

3.5 PARTICLES IN SEAWATER

Particles dispersed in seawater play important roles in chemical and biological processes. They remove elements from near-surface waters and transport them to subsurface waters and are food for filter-feeding organisms, which filter large quantities of seawater. Particles also provide a substrate on which other organisms live. (Biological processes are discussed in Chapter 9.)

Most particles (30 to 70%) in seawater come from organisms. Calcium carbonate and silica, from marine organisms, make up 25 to 50% of seawater's particulate matter. A large fraction of biologically derived particles is relatively large, and therefore sink rapidly (hundreds of meters per day), reaching the ocean bottom in a few tens of days. Such short transit times do not permit much degradation by chemical or biological processes. Many of these large particles are **fecal pellets,** produced by animals feeding on microscopic plants growing in near-surface waters. Fecal pellets provide much of the food for bottom-dwelling organisms in the deep ocean.

Seawater also contains large numbers of much smaller particles that sink very slowly. Small particles take years to reach the ocean bottom, allowing ample time for chemical and biological change. These fine particles are brought to the ocean by rivers, by winds blowing over deserts and mountains, and by meteorites falling through the Earth's atmosphere.

The most soluble particles dissolve before reaching the bottom, thus changing the chemical composition of deep-ocean waters. *Their principal effect is to transport nutrients (nitrogen, phosphate, and silicate compounds) necessary for plant growth from near-surface waters to subsurface waters.* (The importance of this process in biological processes is discussed in Chapter 8.)

Some elements are removed from seawater by chemical reactions occurring on particle surfaces. This occurs in sub-surface waters for such elements as iron, lead, and copper. These elements are enriched in deep-ocean sediments, especially manganese nodules or crusts that form on bare rock surfaces. (We discuss this more in Chapter 9.)

3.6   PROCESSES CONTROLLING SEAWATER COMPOSITION

Salts dissolved in seawater come from three sources: volcanic eruptions; chemical reactions between seawater and hot, newly formed oceanic rocks; and the weathering of rocks on land (Figure 3.7). Sea salt composition has remained nearly constant over hundreds of millions—probably billions—of years, controlled by chemical and biological reactions.

Volcanic eruptions release large volumes of gases, which eventually reach the ocean. These eruptions are how both sulfate and chloride in seawater are primarily derived.

Chemical interactions between seawater and recently formed oceanic crust are another factor controlling sea salt composition. Magnesium and sulfate are removed while other elements are added, like rubidium and lithium. The circulation of the entire volume of ocean water through the ocean crust takes 5 to 10 million years and is probably the primary control that has kept the composition of seawater constant over billions of years.

Many salts in seawater also come from weathering of rocks on land. As rocks decompose to form soils, they release various constituents, among them the soluble constituents like silica, carried to the ocean by rivers, along with

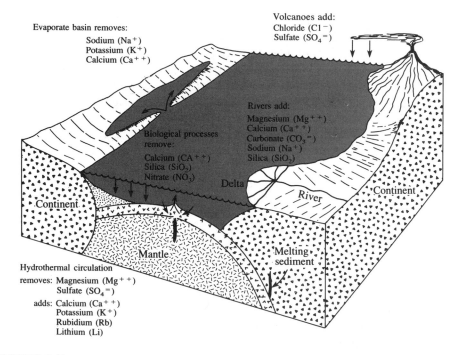

FIGURE 3.7
Schematic representation of processes controlling the composition of sea salts.

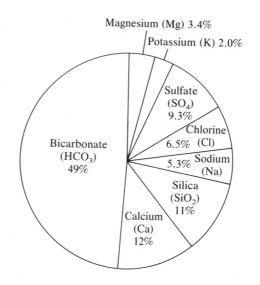

FIGURE 3.8
Relative abundance of dissolved solids in river water.

various metals: sodium, calcium, and magnesium. River waters also contain calcium and bicarbonate from the breakdown of limestone (Figure 3.8). Upon entering the ocean, dissolved salts remain behind while water itself continues to move through the hydrological cycle (discussed later in this chapter).

Since we know that seawater has retained roughly the same salinity over billions of years, there must be processes removing dissolved constituents from seawater. Some constituents are removed by chemical reactions with weathered minerals in sediment particles. Others—such as silica and nitrate—are removed by growing organisms. Still other constituents, such as sodium and chloride, are removed when salt beds form during evaporation of seawater in isolated arms of the sea. Ancient sediment deposits around the world include thick beds of salt, which are the remains of ancient seas that completely dried up, usually many times. Such evaporite deposits are especially likely to form during the early stages of ocean basin formation.

3.7   THERMAL PROPERTIES AND CHANGES OF STATE

Changes in a substance's physical form, such as a solid changing to a liquid or to a vapor, are called changes of state. A change of state involves breaking intermolecular bonds or forming new ones. If bonds are broken, energy is taken up. When new bonds are formed, energy is released. This energy is normally supplied or released as heat.

A calorie (abbreviated **cal***) is the amount of heat (or energy) required to raise the temperature of 1 g of liquid water by 1 °C.* For example, 100 cal must

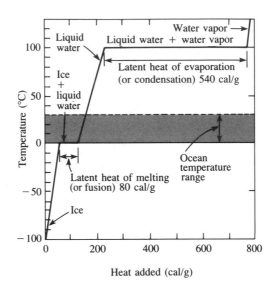

FIGURE 3.9
Changes in temperature when heat is added to (or removed from) ice, liquid water, or water vapor. Note that the temperature of the system does not change when mixtures of ice and liquid water or liquid water and water vapor occur together.

be supplied to 1 g of water to change its temperature from the melting point (0 °C) to the boiling point (100 °C).

To illustrate relationships between addition of heat and changes in temperature as well as changes of state in our water system, let us take 1 g of ice and heat it (Figure 3.9). Water molecules in ice vibrate because molecular bonds are somewhat elastic. Energy supplied by heating causes these molecular vibrations to become stronger and the molecular bonds to stretch.

At the melting point of ice (0 °C), molecular vibrations are strong enough to break molecules loose from the ice structure, forming liquid water. If we momentarily stop adding heat, ice and liquid water can exist together in **equilibrium**. At equilibrium the number of molecules gaining enough energy to break free of the ice structure at any instant is balanced by the number of water molecules losing energy and rejoining the ice structure. Unless we add or remove energy, the amount of ice and liquid water remains fixed.

Adding more heat causes more ice to melt. Removing heat (cooling) has the opposite effect. As long as ice and water exist together, adding more heat does not change the system's temperature. Instead, the added energy breaks more bonds in the ice structure, causing melting. After adding enough heat (about 80 cal/g), the last bit of ice disappears. The amount of heat necessary to melt 1 g of ice at 0 °C is called the **latent heat of melting**. The heat consumed is released when water freezes, hence the term—latent heat.

After the last bit of ice disappears, additional heating causes the temperature to rise. Energy, no longer entirely used up in breaking bonds, causes molecules to move more rapidly, and more rapid molecular movement causes a rise in temperature. Between the melting and boiling points, addition of a fixed amount of heat causes a nearly constant rise in temperature. (The same situation would exist if we cooled water vapor rather than heated ice.)

The relationship between the amount of heat supplied and the resulting temperature change is called **heat capacity**. (This relationship was used to define the calorie.) Several times as much heat must be supplied to water to cause a 1 °C rise in temperature than is necessary to cause a 1 °C rise in temperature of an equal mass of granite or basaltic rock. (You may notice this on a summer day at the beach. The ground becomes hot during the day and cools quickly at night, whereas water temperatures change very little over a 24-hour period.)

Liquid water's heat capacity is unusually high for a liquid because of water's structure. Therefore, *ocean water can absorb or release large amounts of heat and yet change its temperature very little.*

When liquid water reaches 100 °C, its boiling point at normal atmospheric pressure, many molecules have enough energy to break free, forming water vapor, a gas. A relatively large amount of heat energy (539 cal/g) is needed to evaporate water at 100 °C. The reason for this is easily understood. When water evaporates, hydrogen-bonded clusters are broken up. Because of the strength of the hydrogen bond, this requires a great deal of energy. Hence water's **latent heat of evaporation** is large.

In contrast, water's latent heat of melting at 0 °C is only 80 cal. Here the structure is not completely destroyed; not all the hydrogen bonds are broken (Figure 3.4). Consequently, melting ice requires much less energy than evaporating water.

Although water boils and forms vapor at 100 °C, water vapor can form from ice or liquid water at much lower temperatures because some molecules gain enough energy to break their bonds and escape. Wet clothes can dry even when completely frozen, because ice sublimates; that is, changes to water vapor without going through a liquid state. Evaporation from the sea surface, which is extremely important to Earth's heat and water budgets, occurs well below the boiling point. The average sea surface temperature is about 18 °C (64 °F).

Evaporation of water below the boiling point requires more heat per gram of water vapor than evaporation at the boiling point. The increase in the latent heat of evaporation reflects the extra work required to break hydrogen bonds at lower temperatures:

Temperature (°C)	Latent Heat of Evaporation (cal/g)
0	595
20	585
100	539

These processes—or changes of state—are reversible. In other words, the latent heat of evaporation can be recovered by condensing water vapor, forming liquid water. Condensing l g of water vapor at 20 °C to form water at the same temperature releases 585 cal.

Evaporation removes heat supplied by the sun from the sea surface. This heat is returned, warming the atmosphere, when the vapor condenses to fall as rain or snow. This heat transport by water vapor accounts for the mild winters of humid coastal areas. Abundant rainfall releases heat in the atmosphere, preventing the much lower winter temperatures found in dry continental regions far from the ocean.

3.8   PHYSICAL PROPERTIES OF SEAWATER

Some properties of seawater change as salt concentrations increase. For instance, changing salinity from 0 °/oo to 40 °/oo causes **viscosity**—internal resistance to flowing—to increase about 5%. Adding sea salts to water also changes the **temperatures of maximum density** and **initial freezing**. Since salt does not fit into the ice crystal structure, it inhibits ice formation and depresses the initial freezing point (Figure 3.10). Adding salt causes the mixture to freeze at temperatures below 0 °C.

Seawater does not freeze completely at a fixed temperature as pure water does. In other words, *seawater has no fixed freezing point.* As seawater freezes, salts are excluded from the ice structure. Consequently, the seawater remaining becomes saltier and therefore freezes at still lower temperatures. Unless cooled to very low temperatures, some concentrated brine remains.

Processes that depress the initial freezing point also depress the temperature of maximum density. Sea salts inhibit development of the clusters that cause volume expansion of pure liquid water near the freezing point. Adding sea salt

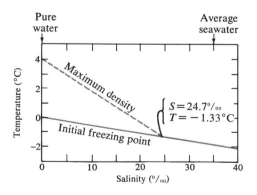

FIGURE 3.10
Effect of salinity on the temperature of maximum density and the initial freezing point of seawater.

to water lowers the temperature of maximum density. At a salinity of 24.7 %oo, the maximum density and the initial freezing temperature occur at − 1.33 °C; at salinities greater than 24.7 %oo, seawater does not exhibit a density maximum. Hence typical seawater (salinity of 35 %oo) becomes progressively denser as it cools until it freezes.

3.9 DENSITY OF SEAWATER

Temperature, salinity, and pressure control the **density** *of seawater.* Of the three, temperature and salinity are most important. In the open ocean, seawater density varies only between relatively narrow limits. Consequently, oceanographers must determine seawater density with great precision and must work with very slight differences.

Normally, seawater density is calculated from precise measurements of temperature (accurate to ± 0.002 °C) and salinity (accurate to ± 0.002 %oo) of water samples. From these measurements, density is calculated to one part in 500,000. Such small differences in density are needed to determine the direction and speed of deep-ocean currents, as we shall see in Chapter 5.

Figure 3.11 shows the changes in water density for salinities between 33 %oo and 37 %oo and temperatures from − 3 ° to 30 °C. This encompasses the temperature and salinity ranges of most seawater samples. Figure 3.11 also shows the relative effects of temperature and salinity on seawater density. At 30 °C, a change in salinity from 34 %oo to 35 %oo changes the density from 1.021 to 1.022. Density is changed an equal amount by cooling water with a salinity of 37 %oo from 27.5 ° to 24.3 °C, a change of 3.2 °C. Such temperature changes

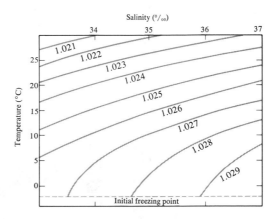

FIGURE 3.11
Changes in the density of seawater (shown in grams per cubic centimeter, g/cm³) caused by changes in salinity and temperature. [After U.S. Naval Oceanographic Office, *Instruction Manual for Oceanographic Observations,* H. O. Pub. 607 (Washington, DC: 1955), p. 42]

occur commonly at the ocean surface. Large salinity changes usually occur near land, due to river discharges and precipitation.

3.10   WATER DENSITY AND STABILITY

The relative density of a particular bit of seawater controls the depth at which that water parcel occurs in the ocean. Changes in density, resulting from processes occurring at the ocean surface, cause the sluggish, deep-ocean currents.

To demonstrate the significance of water density, conduct the following experiments. Fill a glass with tap water (Figure 3.12). Fill a medicine dropper with salty water colored with ink or food coloring. Put a drop of colored salty water into the glass of fresh water. The salty water sinks to the bottom because it is denser than the fresh water. Conversely, a drop of colored fresh water put into a beaker of salty water remains at the surface because it is less dense.

For the next experiment, create a two-layered system. Make dense, salty water by dissolving as much salt as the water in a half-filled container will hold. (Add salt until some remains undissolved even after vigorous stirring.) Then carefully pour tap water on top of the salty water. Wait until the water movements caused by adding the tap water have subsided. Now you have a stable, two-layered system where the denser salty water lies below the fresh water. The system is stable and will remain unchanged unless you mix it by stirring or heating. Stability is resistance to change in the system. If you slightly disturb a stable system, it will return to its initial state after the disturbance ceases.

If you now add slightly salty, colored water to our two-layered system, it comes to rest at an intermediate level. The exact level will depend on the changes of density with depth and the density of the colored salty water. There is a **stable density distribution** if the most dense water (in this case the saltiest)

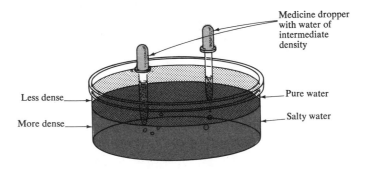

FIGURE 3.12
Vertical movements of intermediate-density water in a stable, two-layered density stratified system.

is at the bottom and the least dense on top. Waters of intermediate density occur at intermediate depths (Figure 3.12).

When a drop of salty water is first added to the system, it momentarily creates an unstable density distribution. In this case, the drop is denser than the surrounding water and it sinks. Conversely, if you put the eye dropper filled with slightly salty water down into the very salty water and release a drop, it rises because you have again created an unstable density distribution; the water drop is less dense than the surrounding water. Thus *an unstable system spontaneously moves toward a more stable density distribution.*

You can also make a **neutrally stable distribution** by mixing the water so water density is the same throughout. In fact, if the two-layered system stands long enough, salt will diffuse through the water, eventually resulting in equal water density throughout. In this case the system does not return to its initial state after a disturbance. *A neutrally stable system is easily mixed.*

The effects of temperature variations on water density can be shown by heating one side of a dish with a flame (Figure 3.13). The warmed water is less dense and rises above the heated areas. Cooler water sinks on the other side of the container and flows along the bottom to replace the rising waters. Water movements can be made more visible by adding drops of ink or food coloring.

These density-controlled vertical movements are called **convection**. The resulting water movements are known as **convection currents**. Convection is especially familiar in the atmosphere, which is warmed at the bottom and cooled at the top. Ocean waters, however, are heated and cooled at the top. Consequently, convective currents are relatively rare in the ocean, occurring in a few locations.

The situation in the ocean can be made experimentally by shining a heat lamp on the surface of a dish of cool water. This creates a layer of warm, less dense water on top, a stable density distribution. Vigorous stirring is required to mix such a system.

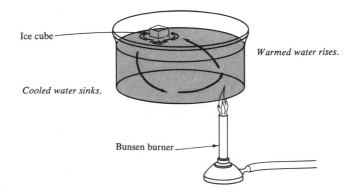

FIGURE 3.13
Convection caused by warming the water at the bottom and cooling it at the top.

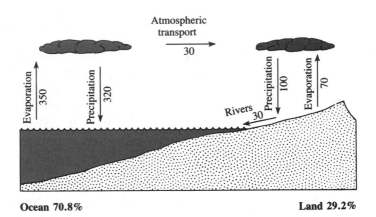

FIGURE 3.14
Schematic representation of the hydrologic cycle, showing the movements of water (in thousands of cubic kilometers per year) from the ocean surface to the continents and back to the ocean by rivers.

3.11   HYDROLOGIC CYCLE

Water vapor constantly evaporates from the ocean surface, leaving behind sea salts. Most water (about 91%) returns to the ocean surface as rain (Figure 3.14). The remaining 9% is carried by winds to fall as rain or snow on the land. Most of that precipitation (about 70%) also evaporates. About 30% of precipitation on land returns to the sea through river runoff. This is called the hydrologic cycle.

QUESTIONS

1. Using water as an example, describe the three forms of matter.
2. Describe the atomic structure of water. Why is water called a polar molecule?
3. How much heat (in calories) must be removed per square centimeter of water surface to cool by 1 °C a water column 100 meters thick?
4. Explain why ice is less dense than liquid water.
5. Define latent heat. Why is the latent heat of evaporation of water greater than the latent heat of melting?
6. Explain why fresh water is densest at 4 °C.
7. How much heat (in calories) is required to bring 5 grams of ice at 0 °C to liquid water 5 °C?
8. Why does seawater not freeze completely at a fixed temperature?
9. List, in order of abundance, the six most common constituents in seawater.
10. What three factors control seawater density?
11. Explain why carbon dioxide makes up 15% of the gases dissolved in seawater but only 0.03% of the gases in the atmosphere.
12. Why are the dissolved-oxygen concentrations in seawater considered to be non-conservative properties?
13. Define salinity.

14. How does circulation of seawater through newly formed oceanic crust alter the composition of seawater?

SUPPLEMENTARY READING

Books

Davis, Kenneth S., and Day, John Arthur. *Water: The Mirror of Science*. Garden City: Anchor Books, Doubleday and Company, Inc., 1961. Elementary, nontechnical.
Deming, H. G. *Water: The Foundation of Opportunity*. New York: Oxford University Press, 1975. Elementary.
Horne, R. A. *Marine Chemistry*. New York: Wiley Interscience, 1969. Technical, good bibliography.

Articles

Gabianelli, J. J. "Water—the Fluid of Life." *Sea Frontiers* 62(5):258–270.
MacIntyre, Ferran. "Why the Sea is Salt." *Scientific American* 223(5):104–115.
Revelle, R. "Water." *Scientific American* 209(3):93–108.

KEY TERMS AND CONCEPTS

Polar molecule
Forms of matter:
  solid,
  liquid,
  gas
Calorie
Heat capacity
Equilibrium
Latent heat:
  melting,
  evaporation

Density distributions:
  stable,
  unstable,
  neutrally stable
Convection
Sea salt composition
Ionic bonds
Hydrogen bonds

Temperature of
  maximum density
Temperature of initial
  freezing
Dissolved gases
Nonconservative
  properties
Conservative properties
Sea ice
Hydrologic cycle
Salinity

4

Open Ocean

In the **open ocean,** oceanic processes are dominated by incoming solar energy and by winds. Here we first look at the effects of the processes we have been discussing in previous chapters. In Chapter 7, we will consider the **coastal ocean,** where processes are complicated by river discharge, complex shorelines, and tides.

4.1   LAYERED STRUCTURE

Light, temperature, and salinity control the behavior of ocean waters. Their distribution results from absorption of incoming solar radiation (**insolation**) in surface waters and transport of heat and water vapor over the Earth's surface. Part of the sun's energy heats the ocean surface, which in turn supplies energy to drive the winds. Much of the sun's energy is used to evaporate water. The atmosphere and ocean together form an inefficient, sun-powered engine that changes a small fraction of the incoming solar energy into winds or ocean currents.

Absorption of insolation at the ocean surface causes a three-layered structure of the open ocean: the **surface, pycnocline,** and **deep zones**. The surface zone changes because of seasonal variations in heating, cooling, evaporation, and precipitation. In the polar and subpolar oceans, freezing of surface seawater to form sea ice is important. These processes control the temperature, salinity, and therefore the density, of surface waters.

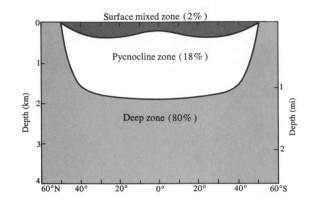

FIGURE 4.1
Schematic representation of the ocean's layered structure, showing the relative volume of each zone. Note that the deep zone comes to the surface near the poles in the Northern and Southern hemispheres. Also note that the pycnocline comes closer to the surface near the equator.

The **surface zone** contains less dense water, usually as a result of higher temperatures caused by warming of surface waters. The thickness of the surface zone reflects the depth of mixing, caused primarily by winds. In certain areas, convective vertical water movements are caused by density changes resulting from changes in temperature and salinity. Extensive mixing of water within the surface zone results in its nearly neutral stability, allowing water particles to move vertically. This zone is also frequently called the mixed layer. Surface waters have ample opportunity to adjust their temperatures to local conditions.

Below the surface zone is the **pycnocline** (**pycno** = density; **cline** = slope), where water density changes markedly with depth. Because of these large changes in density with depth, water in the pycnocline zone is very stable. The pycnocline acts as a barrier to vertical water movements and serves as a floor to the surface circulation with its seasonal temperature and salinity changes.

In addition, the pycnocline acts as a ceiling for the **deep zone** and prevents deep-ocean waters from readily mixing with surface waters or equilibrating with the atmosphere. Only in high latitudes and in polar areas where the pycnocline is usually absent (Figure 4.1) are deep waters exposed to the atmosphere and able to exchange gases. Waters below the pycnocline have an average temperature of only 3.5 °C (38 °F).

4.2   WATER MASSES

Large volumes of seawater move through the oceans as discrete **water masses,** identifiable by their characteristic temperatures and salinities. Water masses form at the ocean surface, and their temperatures and salinities reflect surface conditions. If the water mass is denser than surrounding waters, it sinks to a level determined by its density and the density distribution in the nearby ocean. Below the surface, water masses move with the subsurface currents, often for thousands of kilometers. After hundreds or thousands of years, the waters return to the surface to exchange gases with the atmosphere and to be warmed by heat from the sun. Using changes in dissolved gas concentrations, oceanographers trace subsurface water mass movements.

The densest water masses in the ocean form where waters of relatively high salinity are intensely cooled at the ocean surface (usually in polar regions). These processes increase the depth of the pycnocline by the sinking of dense waters from the surface. If they are dense enough, subsurface water masses may reach the bottom and flow along the ocean floor. If not, a water mass will flow at a level appropriate to its density between the denser bottom waters and the less dense waters of the surface zone. Movements of a water mass of intermediate density are like moving a card in a deck of cards. In the ocean, the layers below have greater densities and the ones above have lower densities.

4.3 TEMPERATURE: HEATING AND COOLING

Heating of the ocean surface occurs during daylight, and it is warmest in late afternoon. The amount of energy the ocean absorbs depends on local cloud cover and the sun's altitude. The sun's altitude depends in turn on latitude and the time of year (Figure 4.2). More energy is absorbed when the sun is high in the sky and less is absorbed when the sun is near the horizon. In the tropics and subtropics the sun is well above the horizon at all seasons. Near the poles the sun is never far above the horizon, and the polar and subpolar regions receive much less insolation. Consequently, *the Earth is heated by the sun in the tropics and subtropics and cooled by radiating energy, primarily from polar and subpolar regions* (Figure 4.3).

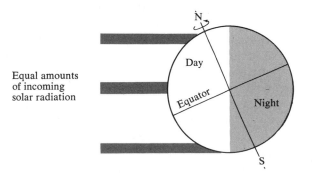

FIGURE 4.2
Variations in incoming solar radiation per unit area result from variations in the angle at which the sun's rays strike the Earth's surface. Conditions are shown for northern summer.

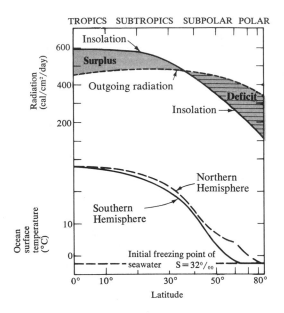

FIGURE 4.3
Radiation balance in the Northern Hemisphere and average surface ocean temperatures
in the Northern and Southern hemispheres. [Heat budget after H.G. Houghton, "On the
Annual Heat Balance of the Northern Hemisphere," *Journal of Meteorology* 11 (1954),
p. 7; ocean surface temperatures from W.E. Forsythe, (Ed.), *Smithsonian Physical Tables,*
9th rev. ed. (Washington, D.C.: Smithsonian Institution, 1964), p. 726]

Energy received from the sun at the top of the atmosphere is relatively
constant, averaging about 0.5 cal/cm² of the Earth's surface per minute. After
passing through the atmosphere, insolation at the Earth's surface is 0.25 cal/cm²
per minute, averaged over 24 hours.

If the average insolation remained in the upper 1 m of the surface zone,
water temperatures would increase about 3.5 °C in one day. However, the
observed average daily temperature variation in the open ocean is only 0.2 °
to 0.3 °C. This means that the daily input of solar energy is quickly mixed
through near-surface waters or lost through evaporation to the overlying atmosphere. Since heat gained by the ocean during the day is distributed through
a fairly thick surface zone, it is not readily lost at night. Thus this mixing, combined with water's relatively high heat capacity, prevents a large daily change
in surface water temperatures. On land, heat remains near the surface (there
is little conduction) during the day and is therefore readily lost at night. For
this reason, daily temperature ranges on land are much greater than temperature
ranges over the ocean.

If the ocean retained all the heat it absorbed, ocean waters would reach
the boiling point in less than 300 years. Obviously, this is not happening.
Furthermore, fossil remains of ancient marine organisms preserved in rocks

TABLE 4.1
Heat budget of the ocean surface (24-hour average)

		Heating	Cooling
		(cal/cm²/min)	
Incoming solar radiation		0.25	
Radiation back to space			0.10
Evaporation			0.13
Atmospheric warming			0.02
	Totals	0.25	0.25

show that ocean surface temperatures have changed little in the past 1 to 2 billion years. This means that *the ocean loses as much energy as it absorbs from insolation* (Table 4.1).

Heat loss from the ocean surface continues day and night and in all seasons. Three processes are involved: (1) radiation of heat back to space; (2) heating of the atmosphere by conduction; and (3) evaporation of water (Table 4.1). About 40% of the insolation received by the oceans radiates back to space (Figure 4.4). Most of this is in the infrared part of the spectrum instead of the visible spectrum emitted by the much hotter surface of the sun.

Some of the ocean's heat loss goes directly into warming the atmosphere. Heat passes to the atmosphere by conduction, just as heat is passed on to a pan on a hot stove. Usually the ocean surface is about 1 °C warmer than the overlying air.

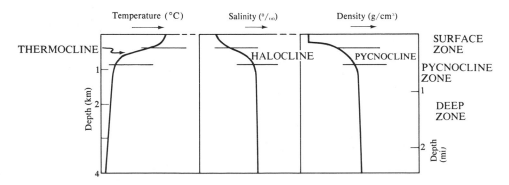

FIGURE 4.4
Typical vertical distribution of temperature, salinity, and density in the ocean. [After J.P. Tully, "Oceanographic Regions and Assessment of Temperature Structure in the Seasonal Zone of the North Pacific Ocean," *Journal of Fisheries Research Board* 21 (Canada: 1964), p. 942]

About half the ocean's heat loss is due to evaporating water. This heat is given off to the atmosphere when water vapor condenses to fall as rain or snow. Much of the atmosphere's heat is gained in this way.

One result of heating the ocean's surface is the three- layered structure already described. Rapid vertical transfer of heat in the surface zone creates a nearly isothermal (iso = equal; thermal = heat) mixed surface zone. Separating the sun-warmed surface zone and the cold, deep zone is the thermocline (thermo = heat; cline = slope), where temperature changes abruptly with depth (Figure 4.4). Below the thermocline, temperatures change little with increasing depth. Changes in temperature cause density changes, so *in many ocean areas the thermocline coincides with the pycnocline.*

Unequal heating of the Earth causes large differences in surface water temperatures between tropical and polar regions (Figure 4.5). The ocean is warmest (25° to 30°C) in the tropical and subtropical regions and coldest (down to −1.7°C) near the poles (Figure 4.3). In general, *belts of equal surface-water temperature run east-west.*

Surface isotherms—lines connecting areas of equal temperature—deviate from their east-west trends near continents. This is especially obvious in the western portions of the North Atlantic and North Pacific oceans. These deviations are caused by the continents and by ocean boundary currents, which tend to parallel the shorelines. Some boundary currents, such as the Gulf Stream, transport warm water toward the poles. Others, like the Labrador Current, transport cool water toward the equator.

Comparison of ocean surface temperatures (Figure 4.3) in the Northern and Southern hemispheres shows that temperature changes with latitude are similar in both hemispheres. The annual differences in surface temperature are greatest in the subpolar oceans, around 60° North and South latitudes, and in the land-dominated belt of the Northern Hemisphere, and are least in the ocean-dominated belt of the Southern Hemisphere.

The lowest surface water temperature (−1.7°C) coincides with the temperature of initial freezing (Figure 3.9) for seawater with a salinity of about 32°/oo. Freezing or melting of sea ice in the polar oceans acts as a thermostat, essentially fixing surface water temperatures. Local surface water temperature cannot go higher until sea ice melts, and as long as some surface water remains unfrozen at the same salinity, water temperatures cannot go lower.

4.4   SALINITY: EVAPORATION AND PRECIPITATION

Salinity in the open ocean varies much less than temperature (Figure 4.6). *Changes in salinity are caused primarily by evaporation (removal of fresh water as water vapor), by precipitation (adding fresh water as rain or snow), and by river discharge.*

In the high latitudes, sea-ice formation plays an important role because nearly fresh water goes into the ice, leaving behind the salts (see Chapter 3).

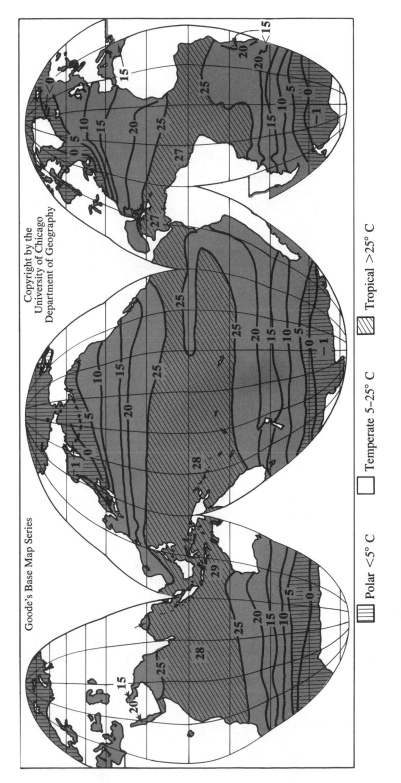

FIGURE 4.5

Ocean surface temperatures (°C) in northern winter. Note the east-west trends of the temperature zones. [After H.J. McLellan, *Elements of Physical Oceanography* (Oxford: Pergamon Press, Ltd., 1965) p. 44]

Polar <5° C   Temperate 5–25° C   Tropical >25° C

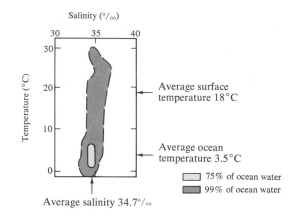

Salinity (°/₀₀)

FIGURE 4.6
Range of temperature and salinity in the world ocean. [After R.B. Montgomery, "Water Characteristics of Atlantic Ocean and of World Ocean," *Deep Sea Research* 5 (1958), p. 144]

These processes act on the ocean surface along with heating and cooling processes (Figure 4.7). The marked changes of salinity with depth (Figure 4.4) form the **halocline (halo** = salt).

Salinity changes affect seawater density. A change in salinity of 1 °/oo causes a greater density change than does a temperature change of 1 °C, which means that in those parts of the ocean where surface waters (Figure 4.7) are diluted by excess precipitation (Figure 4.8), the main pycnocline and halocline frequently coincide. Nevertheless, despite important local effects of reduced surface salinity, over most of the ocean, the pycnocline is controlled by the thermocline. This is primarily a result of the relatively large temperature range (−1.7° to 30 °C) of surface seawaters (Figure 4.6). In contrast, the salinity range for most of the ocean is rather small (33 °/oo to 37 °/oo).

Water evaporated from the ocean surface each year is equivalent to a layer about 1 m thick (39 in). About 90% of this water returns to the ocean surface as rain. The remainder falls as rain (or snow) on the land. Eventually this water also returns via rivers to the coastal ocean, where it causes lower salinities (Figure 4.7).

Evaporation from the ocean surface is controlled by (1) local insolation, (2) wind speed, and (3) relative humidity of the overlying air. Because of the abundant insolation, the tropics and subtropics (Figure 4.3) experience large amounts of evaporation (Figure 4.8). Conversely, in the polar regions evaporation is much less.

Maximum evaporation occurs in subtropical regions (around 30 °N and 30 °S) where the Trade Winds blow throughout the year. The subtropics are also areas of clear skies (high insolation) and relatively dry air, and relatively high surface salinities (greater than 35 °/oo) near 30 °N and 30 °S demonstrate

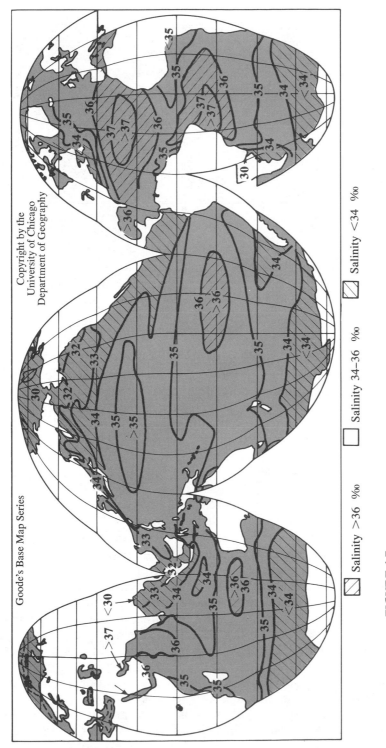

Goode's Base Map Series

Copyright by the
University of Chicago
Department of Geography

FIGURE 4.7

Ocean-surface salinity (in parts per thousand). [After H.U. Sverdrup, M.W. Johnson, and R.H. Fleming, *The Oceans: Their Physics, Chemistry, and General Biology*, © 1970, p.19. Chart VI]. Reprinted by permission of Prentice-Hall, Inc., Englewood Cliffs, NJ.

◪ Salinity >36 ‰   ▢ Salinity 34–36 ‰   ◩ Salinity <34 ‰

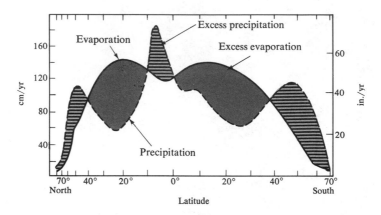

FIGURE 4.8
Distribution of evaporation and precipitation over the ocean. [Data from G. Wüst, W. Brogmus, and E. Noodt. "Die Zonale Verteilung von Salzgehalt, Niederschlag, Verdunstung, Temperatur und Dichte an der Oberfläche der Ozeane," *Kieler Meeresforschungen,* Band V (1954), p. 146]

the excess of evaporation in these areas (Figure 4.8). Diminished evaporation in equatorial regions is due in part to the light and variable winds that give the region its name, the Doldrums. Extensive cloudiness also contributes by diminishing insolation. Precipitation, on the other hand, is most abundant near the equator, in high latitudes, and in coastal regions where surface water salinities are markedly lower (Figure 4.7).

4.5   OCEAN CLIMATE

Climatic regions in the ocean (Figure 4.9) are defined by the similarity of conditions at the ocean surface. The simplest classification separates the surface ocean into coastal (or near-shore) ocean and the open ocean.

Coastal oceans (discussed in Chapter 7), lying over continental shelves, are highly variable. They are influenced by nearness to continents, river discharges, shallowness, and the complexity of nearby shorelines. In contrast, surface waters in the open ocean are more uniform in temperature and salinity. *In the open ocean far from continental boundaries, climatic zones extend nearly east-west across the ocean* (Figure 4.9 and Table 4.2).

Locations of boundaries separating climatic zones are somewhat arbitrary. Boundaries separating climatic regions are marked by **convergences** (except in the tropics), where surface waters flow toward an area and sink below the surface. Convergences in the ocean correspond to weather fronts in the atmosphere, but their positions do not change as rapidly.

Tropical regions straddle the equator in the Atlantic and Pacific oceans. In the Indian Ocean, they lie to the equator's south (Figure 4.9). In the tropics,

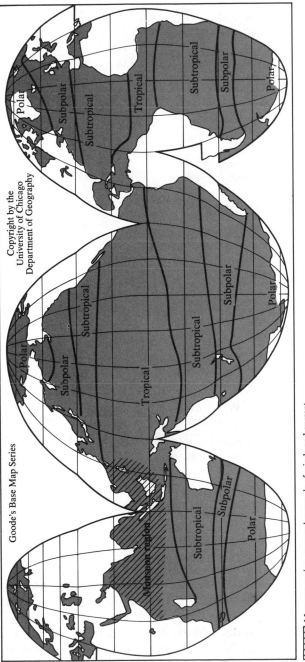

Goode's Base Map Series

Copyright by the
University of Chicago
Department of Geography

 Monsoon region, seasonal reversals of winds and current directions.

FIGURE 4.9
Climatic regions of the ocean.

TABLE 4.2
Open ocean climatic regions

Oceanic Region	Surface Heating and Cooling Cycle	Net Precipitation (+) or Evaporation (−)	Mixing Processes	Seasonal Temperature Range (°C)
Polar	Seasonal melting and freezing of sea ice. Net heat loss.	+	Wind and convective mixing.	Intermediate (5–9)
Subpolar (West Wind drift)	Seasonal heating and cooling. Net heat loss.	+	Wind and convective mixing.	Intermediate (2–8)
Subtropical	Seasonal heating and cooling. Net heat gain.	−	Wind and convective mixing.	Large (6–18)
Tropical	Daily heating and cooling. Net heat gain. No seasons.	+	Wind	Small (•2)

Temperate brackets Subpolar and Subtropical regions.

seasonal temperature changes are slight, and there is a large excess of precipitation over evaporation. Weak and variable winds cause little mixing of surface waters, and the pycnocline is rather shallow.

On either side of the tropical ocean are **subtropical regions** centered around 30°N and 30°S. Trade Winds continually blow across these parts of the ocean. Because of the prevailing winds and abundant sunshine, evaporation exceeds precipitation here. These subtropical areas are major sources of water vapor to the atmosphere.

Seasonal temperature changes are relatively large in subtropical regions, ranging between 6° and 18°C in surface waters. These temperature changes are greatest near the Black Sea. Because of evaporation, subtropical surface waters are more saline and warmer than the average. When they cool during winter, the increased density causes convective mixing. This supplements wind-mixing and is one reason why the regional thermocline is relatively deep under subtropical areas.

Subpolar regions have an excess of precipitation and lie in a belt of strong winds, especially in the Southern Hemisphere. During seasons of high rainfall or large river discharges, a well-developed halocline forms. During local summer, a thermocline usually develops.

In **coastal regions,** river runoff measurably reduces local surface salinities. For example, surface salinities in the subpolar North Pacific Ocean are less than 32°/oo along the coasts.

Polar regions are influenced by seasonal freezing and thawing of sea ice. Some convective mixing in the surface zone usually accompanies sea ice formation. When sea ice forms, highly saline brines are released, which mix with nearby waters.

Near Antarctica, cooling of relatively high-salinity Atlantic waters increases seawater density substantially. The mixing of chilled surface waters with cold, highly saline brines resulting from sea ice formation creates the Antarctic Bottom Water, the densest water mass in the ocean. Since these waters are so dense, they sink to the bottom. After flowing around Antarctica, they spread northward into all three major ocean basins.

4.6   GLACIAL OCEAN

During most of its history, Earth's climate has been warm and humid, with little difference between tropical and polar regions. The past 2 million years, however, have been markedly cooler. Continental glaciers expanded to cover large areas of Northern Hemisphere continents and then retreated to varying degrees about 30 times. Sea level fell to 130 meters below its present level during glacial advances and then rose to be about 50 m higher than present during the glacial retreats. The most recent advance ended about 20,000 years ago. The present climate is warmer than it has been for most of the past 2 million years.

In ocean sediments the distributions and abundances of fossils of one-celled animals, called foraminifera, show how glacial advances affected the ocean. The average sea surface temperature dropped 2° to 3°C, and average land temperatures were about 6.5°C lower than today. At the same time, ocean circulation was more energetic. The Gulf Stream went nearly east from the Carolinas, striking Spain rather than flowing into the Norwegian Sea as it does today (Figure 4.10). When sea level stood about 85 m lower than present, humans migrated from Siberia to Alaska across the exposed Bering Sea floor.

Regular changes in the Earth's orbit around the sun are a major factor controlling the advances and retreats of the glaciers. This theory, advanced by the Yugoslavian mathematician Milutin Milankovitch (1879-1958), predicts climatic changes recurring every 100,000, 40,000, and 10,000 years.

4.7   GREENHOUSE EFFECT

Another major factor in the present warming trend is the changing composition of the Earth's atmosphere. Carbon dioxide released by burning fossil fuels has collected in the atmosphere and ocean, and while the increase is not yet

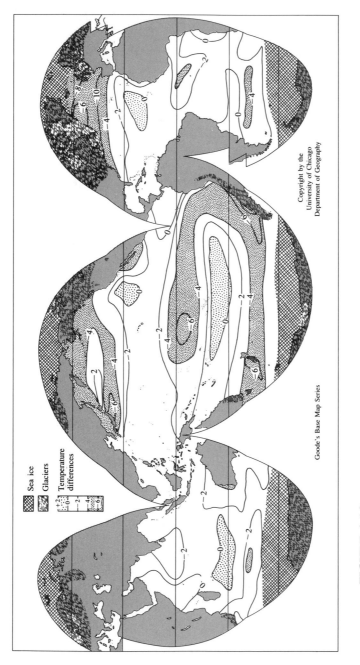

FIGURE 4.10

Changes in sea surface temperatures (August) during the last phase of the Ice Age, about 18,000 years ago. Lower temperatures were widespread at high latitudes and along the equator. Higher temperatures occurred in the centers of the subtropical gyres. [After A. McIntyre, CLIMAP, "The Surface of the Ice-Age Ocean," *Science* 191 (1976) p. 1134]

detectable in the ocean, it has caused atmospheric carbon dioxide levels to rise about 25% above their pre-Industrial Revolution levels. This in turn has caused the **greenhouse effect**—a warming of the Earth's climate similar to the solar warming experienced in greenhouses. As carbon dioxide levels continue rising, temperatures are expected to increase with them. Predictions range from increases of 2.5 to 8 °C in the next fifty years. The effect on the ocean will be most noticeable in the high latitudes where sea ice will be affected. Sea level is expected to rise globally, and this increase will cause flooding of low- lying coastal regions. (This is discussed further in Chapter 10.)

QUESTIONS

1. Explain why plants can produce food only in near-surface waters.
2. What are the average temperature and salinity for the world ocean?
3. Explain why the ocean is heated primarily in the low latitudes and cooled in the high latitudes.
4. List the three processes that cause changes in seawater salinity and indicate where each is most important.
5. List and briefly describe the three major depth zones of the ocean.
6. What three processes cause heat loss from the ocean's surface? Which of these processes has the greatest effect on the atmosphere? Why?
7. What processes cause seasonal variations in surface water properties? Which are most important in the subtropical ocean? Which are most important in the polar ocean?
8. How are oceanic climatic zones related to the atmospheric circulation?

SUPPLEMENTARY READING

Books

Miller, A., and Anthes, R. A. *Meteorology*. 5th ed. Columbus, OH: Merrill Publishing Company, 1985. General reference, elementary.
Perry, A. H., and Walker, J. M. *The Ocean-Atmosphere System*. London: Longman, 1977. Emphasizes ocean-atmosphere interactions.

Articles

Gregg, M. "Microstructure of the Ocean." *Scientific American* 228(2):64–77.
MacIntyre, Ferran. "The Top Millimeter of the Ocean." *Scientific American* 230(5):62–77.
Stewart, R. W. "The Atmosphere and the Ocean." *Scientific American* 221(3):76–105.

KEY TERMS AND CONCEPTS

Heat budget	Layered structure	Precipitation
Pycnocline	Water mass	Climatic zones
Thermocline	Evaporation	Glacial climate
Halocline		

5
Ocean Currents

Ocean waters move unceasingly. Anyone who sails or swims in the ocean knows the horizontal water movements called **currents**. Some currents are transient features and affect only a small area, such as a beach; these are the ocean's response to local—often seasonal—conditions. Other currents extend over large parts of the world ocean; these are the response of the ocean and atmosphere to the energy flow from tropics and subtropics to subpolar and polar regions.

5.1   SURFACE CURRENTS

Knowledge of major surface currents comes principally from compilation of mariners' observations, begun in the 1840s by the American oceanographer Matthew Fontaine Maury. Ships' courses are deflected by surface currents; this deflection causes a discrepancy between a ship's intended position and its actual position after it has traveled for a period. Thus the direction and speed of the local current can be deduced from a ship's actual position after it has been steered on a given course. Combining thousands of such observations, Maury synthesized a generalized picture of the ocean currents. From such data, gathered over many years, charts of average surface currents have been compiled (Figure 5.1).

Changeable currents, such as the seasonal **Monsoon Currents** in the Northern Indian Ocean, are not easily defined by observations gathered in various

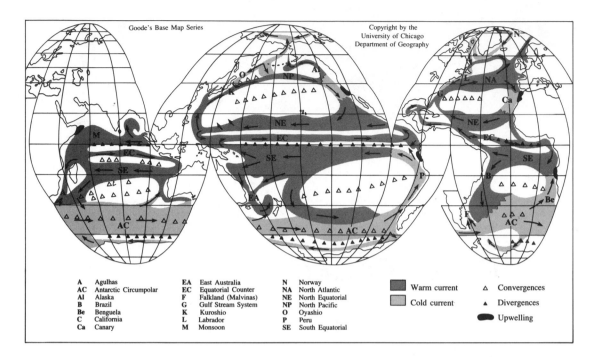

FIGURE 5.1
Ocean-surface currents, January-February. [After Pilot Charts, (Washington, D.C.: U.S. Naval Oceanographic Office), various printings]

seasons over many years. Study of such changeable currents requires observations from a single season. It is difficult to map transient currents; thus we know little about the ocean's short-term variability. Satellite observations are now filling that void.

Near-surface winds and surface currents are closely linked. This can be seen by comparing a generalized picture of prevailing winds with the surface circulation (Figure 5.2). The correspondence is close, except near Antarctica.

Major open-ocean currents form nearly closed current systems called **gyres**. Each ocean basin has a large current gyre in the subtropical regions (around 30°N and 30°S of each hemisphere. Smaller gyres occur in the subpolar oceans, centered near 50°N. In the Southern Ocean, the **Antarctic Circumpolar Current** around Antarctica connects the circulation systems in all three basins.

Each gyre consists of four currents. Open-ocean, east-west currents form the gyre's northern and southern limbs and these are joined by boundary currents sitting nearly parallel to the continental margins, generally oriented north-south.

Open-ocean currents, such as the **North Pacific Current** or the **North** and **South Equatorial currents**, flow at speeds of 3 to 6 km/day (2 to 4 mi/day),

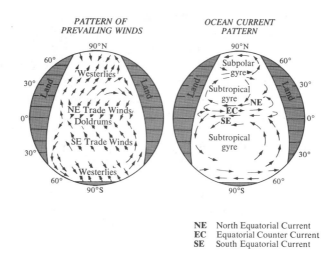

PATTERN OF PREVAILING WINDS

OCEAN CURRENT PATTERN

NE	North Equatorial Current
EC	Equatorial Counter Current
SE	South Equatorial Current

FIGURE 5.2

Generalized atmospheric and surface-ocean circulation. [After R.H. Fleming, "General Features of the Ocean," *Geological Society of America Memoir 67*(1) (1957), p. 95]

and usually extend 100 to 200 m (300 to 650 ft) below the surface. Waters moved by these currents remain in the same climatic zone for many months, giving them ample time to equilibrate with local climatic conditions.

The major eastward-flowing current in the ocean is the Antarctic Circumpolar Current (sometimes called the West Wind Drift). It circles Antarctica, forming part of the Southern Hemisphere current gyres in each basin. The narrow opening between Cape Horn (the tip of South America) and Antarctica (Drake Passage) is a partial barrier to the current. Such a globe-encircling current is possible only around Antarctica. Elsewhere, continents deflect the east-west currents. Part of the water deflected from the Antarctic Circumpolar Currents flows northward to form the **Peru Current** (also sometimes called the Humboldt Current).

Western boundary currents flow generally northward in the Northern Hemisphere and southward in the Southern Hemisphere. These currents are unusually large and swift. The relatively narrow, jet-like currents of the **Gulf Stream** system and the **Kuroshio** off Japan, are the largest currents in the ocean. They have speeds between 40 and 120 km/day (25 to 75 mi/day), and their flows extend much deeper below the surface than other currents, down to depths of 1000 m (3300 ft) or more. The flow of each of these currents is 50 to 100 times the flow of all the world's rivers combined.

Because of their rapid flow, surface waters in the western boundary currents do not adjust completely to local climatic conditions. As a result, boundary currents transfer large amounts of heat from the tropics toward the poles. Western boundary currents of the Southern Hemisphere, such as the Brazil Current and the **East Australia Current,** are not as prominent as those in the northern oceans. This is in part because land is not such a barrier to ocean flow as it is in the Northern Hemisphere.

Eastern Boundary Currents, such as the **California Currents** and the **Canary Current,** are slower and broader than their western counterparts. The generally north-south flow of eastern boundary currents takes surface waters across climatic zones rather slowly, from 3 to 7 km/day. This permits surface waters to adjust at least partially to local conditions. By transporting colder water toward the tropics, eastern boundary currents, such as the Peru and Benguela currents, also transport heat. (They are like the return flow in a household heating system.)

Trade Wind belts border the tropics on the north and south (Figure 5.2). These highly persistent winds drive the North and South Equatorial currents westward. The Equatorial Counter Current is the partial return (eastward) of waters carried westward by the North and South Equatorial currents. This return flow occurs near the equator.

The Southeast Trade Winds of the Southern Hemisphere extend across the equator into the Northern Hemisphere. The **Doldrums,** a region of light and variable winds separating the Trade Wind systems of the Northern and Southern hemispheres, generally lie north of the equator. Consequently the eastward flowing **Equatorial Counter Current,** separating the current systems of the two hemispheres, also lies just north of the equator. The South Equatorial Current crosses the equator in the Atlantic and to a lesser extent in the Pacific. In this way, it transports surface waters into the Northern Hemisphere. The return flow is through subsurface currents that we discuss later.

5.2   CORIOLIS EFFECT

Surface waters, sea ice, and any other substance set in motion by winds move obliquely to the right of the wind in the Northern Hemisphere and to the left in the Southern Hemisphere (Figure 5.3). Caused by the Earth's rotation toward the east, this deflection of objects moving long distances over the Earth markedly affects directions of both currents and winds as seen from the Earth's surface.

Let us see what causes this. A particle moving from the equator toward the pole also moves eastward at about 1670 km/hr (about 1050 mi/hr) because of the Earth's eastward rotation. After moving northward to latitude 30°N (about the latitude of New Orleans), the same particle not attached to the Earth's surface is still moving eastward at 1670 km/hr. But at New Orleans, where the latitude line around the Earth is shorter, the Earth's surface moves eastward more slowly, at about 1450 km/hr (935 mi/hr). The particle is now moving eastward faster than the Earth's surface at that point. To an observer on the ground at New Orleans following the particle's movements, it seems to have been deflected eastward (to the right of its original path). An observer on the moon, however,

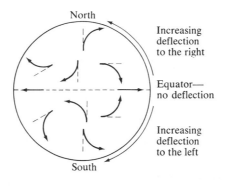

FIGURE 5.3
Paths of moving objects are deflected by the Coriolis effect. There is no deflection at the equator; the deflection increases toward the poles. [After A.N. Strahler, *Physical Geography* 2d ed. (New York: John Wiley & Sons, Inc., 1960) p. 129]

could see that the particle actually moved in a straight line, but the Earth's surface at different latitudes moved eastward at different speeds beneath it.

This apparent deflection of a moving particle, to the right in the Northern Hemisphere and to the left in the Southern Hemisphere, is called the **Coriolis effect**. The amount of deflection depends on the latitude and on the particle's speed. The Coriolis effect is zero for a particle moving along the equator and is a maximum at the poles. Particles at rest are unaffected. The Coriolis effect is important when other forces acting on a moving particle are small and when the particle has moved a long distance. Your car is not deflected by the Coriolis effect because it is in contact with the ground.

5.3 EKMAN SPIRAL

Winds blowing steadily across water move it by dragging on the surface. Wind ripples or waves cause the surface roughness necessary for the wind to "grab" surface waters and set them in motion. A steady wind blowing for 12 hours with an average speed of 100 cm/sec (about 2.2 mi/hr) over deep water causes surface currents of approximately 2 cm/sec. In other words, the resulting current has a speed that is about 2% of the wind that set it in motion.

Although winds cause surface waters to move, the resulting currents commonly involve waters down to 100 m (300 ft) below the surface. This is because slowly moving fluids flow as thin sheets sliding over each other, called **laminar flow**. Because of viscosity, each moving layer drags on deeper, slower moving ones. At these slow speeds, movement is transmitted from one layer to another by collisions of individual water particles. Energy is transferred from rapidly moving layers to slower moving ones and is steadily lost in overcoming **molecular viscosity** (caused by molecular interactions) within each layer. As a result, wind-induced surface currents are not transmitted to great depths in the ocean.

Ocean waters commonly move too rapidly for laminar flow to persist. Instead, currents are usually **turbulent**. Water particles move in irregular, ever-changing eddies that are carried along by the main flow. Eddies interact, transferring motions from one to another. Resistance to flow resulting from these interactions is called eddy viscosity. Eddies transfer momentum between themselves many thousand times more rapidly than energy transfers between molecules. Eddy viscosity varies greatly, depending on density stratification and speed of flow.

A steady wind blowing across an infinite, homogeneous ocean with uniform eddy viscosity in the Northern Hemisphere causes surface waters to move at an angle of 45° to the right of the wind (45° to the left in the Southern Hemisphere). Each layer of moving water sets the layer below in motion. As each deeper layer is set in motion, it is also deflected by the Coriolis effect, causing it to move to the right of the overlying layer. Deeper layers move more slowly because energy is lost in each transfer between layers.

We can plot movements of each layer using arrows whose length represents the speed of movement and whose direction corresponds to the direction of the layer's movements. The idealized pattern for a surface current set in motion by the wind in the Northern Hemisphere (Figure 5.4) is called an **Ekman Spiral**, named after V. W. Ekman, a German physicist who first explained the relationship between winds and surface currents. Each layer is deflected to the right of the overlying layer, so the direction of water movements shifts, increasing depth. At some depths, often around 100 m (about 300 ft), water moves slowly in a direction opposite to the surface layer. This is usually considered to be the base of the wind-driven currents. Combining movements in all layers in the Ekman Spiral, we find that *net water movement is perpen-*

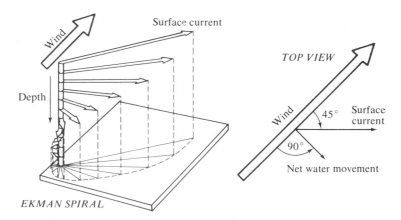

FIGURE 5.4
Water movements in a wind-generated current in the Northern Hemisphere.

dicular to wind direction (Figure 5.4), *90° to the right of the wind in the Northern Hemisphere (90° to the left in the Southern Hemisphere).* Such wind-induced water transport is important in surface circulation, especially near coasts.

This model is an idealized case, involving an infinite, homogeneous ocean (no pycnocline and no boundaries). Since the ocean does not satisfy these conditions, actual wind-induced water movements often differ appreciably from these predictions. For example, the angle between the directions of winds and surface-water movements varies from 15° in shallow waters to the theoretical maximum of 45° in the deep ocean. The pycnocline also inhibits energy transfers to deeper waters, in most cases confining the wind-driven currents to the surface zone. The pycnocline is the floor of the surface currents.

5.4   GEOSTROPHIC CURRENTS

Let us examine large currents caused by prevailing winds (Figure 5.2). Remember that in deep waters the net water movement (Northern Hemisphere) is 90° to the right of the wind (Figure 5.4). Thus prevailing winds move surface waters toward the center of the basin (Figure 5.5).

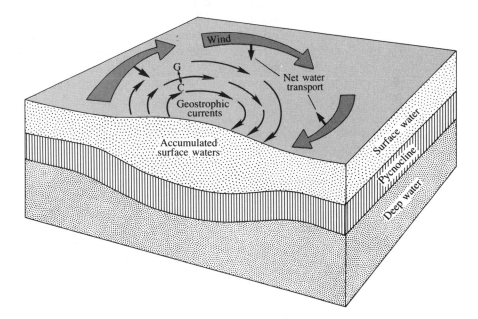

FIGURE 5.5
Net water movements resulting from a nearly closed wind system in the Northern Hemisphere cause surface waters to accumulate (converge), forming a low hill of water. Geostrophic currents result from sloping sea surfaces and the balance between the effect of gravity as the water flows downhill and the deflection (to the right in the Northern Hemisphere) due to the Coriolis effect.

When waters flow toward an area, it is called a **convergence**. Wind-induced convergences form low hills of less dense water near the center of ocean basins since prevailing winds tend to blow around the basin margins. The resulting sea surface slopes are very gentle, only one to two meters (usually less) above an arbitrary level over thousands of kilometers.

The steepest slopes occur on the western side of ocean basins where the currents are strongest. These slopes cannot be measured directly, as we might survey a hill on land. Instead, oceanographers calculate "**dynamic topography**" using measurements of temperature and salinity. Let's see how that works.

Remember that less dense water occupies more volume than dense water. This means that two columns of water of equal mass but with different densities will have different surface elevations above a level bottom. The less dense column will stand higher than the denser column. If we can determine the density of various water columns we can calculate the slopes of the water surface between stations. By making many determinations, it is possible to map sea-surface slopes over a region or basin.

Knowing this surface topography, it is possible to calculate speeds and directions of surface currents. Water responds to a sloping ocean surface by flowing downhill. The water starts to move downhill but its path is deflected by the Coriolis effect to the right in the Northern Hemisphere. The water changes flow direction until the force of gravity is balanced by the Coriolis effect. *In an idealized, frictionless ocean, gravity acting in a downhill direction is balanced by the Coriolis effect acting to the right (uphill in an equilibrium situation), resulting in a geostrophic (Earth-turned) current.* In the Northern Hemisphere, the hill of less dense water is on the right when one looks in the direction of flow; in the Southern Hemisphere it is on the left.

The major ocean currents are geostrophic. Thus currents can be determined by charting sea surface topography. Currents flow around elevations or depressions; the steepness of the slopes controls current speed. *Current speeds are greater on a steep slope and less on a gentle slope.*

In an idealized ocean, where water had no viscosity, currents would flow at a constant elevation around the hill of water, never reaching the bottom of the hill. However, seawater is somewhat viscous and energy must be expended to keep water flowing downhill so that it will eventually reach the bottom. Geostrophic currents are simplified representations of a more complex world, but they are useful to predict tracks of icebergs moved by currents or floating debris from wrecked ships or aircraft.

5.5   GULF STREAM RINGS

Ocean-sensing satellites show a more dynamic ocean than originally suspected. Early oceanographers spoke of currents as "rivers in the sea," implying fixed locations for them. This image of oceanic constancy has been replaced by a picture of ocean currents almost as changeable as the winds.

Among the most dynamic currents are rings spun off by Western Boundary currents, especially the Gulf Stream, the Kuroshio, and the Eastern Australian Current. Such **rings** begin as meanders which grow and eventually break off to move independent of the main current (Figure 5.6). Strong currents surround rings, isolating waters and organisms in the rings from the surrounding waters. Rings are most common in the western portions of ocean basins (Figure 5.7).

Rings form on both sides of the Gulf Stream. Those forming on the north side enclose parcels of warm Sargasso Sea water 100–200 km (60–100 mi) wide lying to the south of the Gulf Stream; these are called **warm core rings**. Since the rings are up to 2 km deep, they cannot easily come up on continental shelves.

Rings on the south of the Gulf Stream contain cold, low-salinity waters from the coastal ocean. These are called **cold core rings**. Rings on both sides

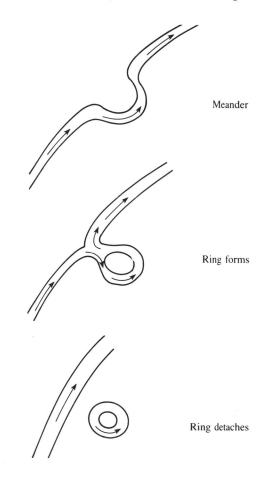

Meander

Ring forms

Ring detaches

FIGURE 5.6
A ring forms from a meander in a western boundary current. After separating, the ring moves slowly in the direction opposite to the boundary current.

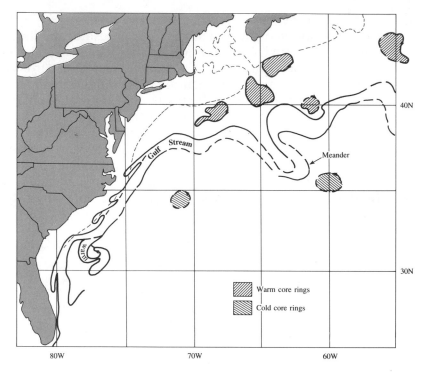

FIGURE 5.7
Rings are common along the U.S. Atlantic coast.

of the Gulf Stream move slowly (5–6 km or 3–5 mi/day) southeastward. Warm core rings are reabsorbed into the Gulf Stream after a few months to a year. Cold core rings last up to three years. These systems are similar to storms in the atmosphere.

5.6   WIND-INDUCED VERTICAL WATER MOVEMENTS

Winds cause vertical water movements. Both upward (upwelling) and downward (sinking) water movements can be caused by winds blowing across the ocean surface. Coastal upwelling and sinking occur where prevailing winds blow parallel to the coast. Winds cause surface waters to move but the presence of land or a shallow bottom restricts water movements (Figure 5.8). When the net wind-induced water movements are offshore, subsurface waters flow to the surface near the coast. This slow, upward flow, from 100 to 200 m (300 to 650 ft) deep, replaces surface waters blown seaward.

Coastal upwelling is common along the western coasts of all continents. Cool summer weather with frequent coastal fogs results chiefly from the

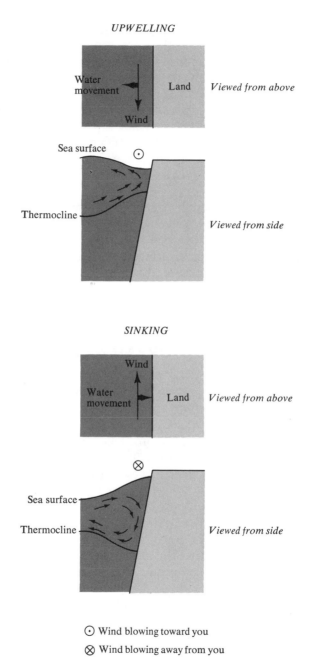

FIGURE 5.8
Wind-induced coastal upwelling and sinking in the Northern Hemisphere. Slopes of the sea surface and pycnocline are greatly exaggerated. The arrows show directions of water movements.

presence of this cooler, upwelled, subsurface water. Upwelled waters are iden-tifiable by their low dissolved-oxygen contents since they have not recently been in contact with the atmosphere. Vertical water movements bring to the surface waters rich in dissolved nutrients (nitrogen and phosphate compounds) to sup-port abundant growth of phytoplankton. Thus, *areas of upwelling commonly support areas of major fish production.* About half the world's fish production comes from upwelling areas.

Upwelling also occurs in the equatorial open ocean. This wind-induced upwelling is caused by the change in direction of the Coriolis effect at the equator. Westward-flowing, wind-driven surface currents near the equator flow northward on the north side and southward on the south side of the equator. This process is called **divergence**. Surface waters moving away from the equator cause upwelling of deeper waters.

Downward movements of coastal waters (sinking) occur when the wind-induced movement of surface waters is shoreward. Effects of coastal sinking are less obvious to coastal dwellers than the effects of upwelling, although the local abundance and distribution of fish may be changed radically by sinking surface waters.

5.7   WATER MASS MOVEMENTS

Cold water masses in the deep-ocean basins form primarily in the polar regions, especially near Antarctica. There, surface waters are strongly cooled and sea ice forms. Salts excluded during freezing mix with the already chilled water, and this pronounced cooling and increased salinity form cold, dense waters that sink to the ocean bottom. Once on the bottom, these dense waters flow around Antarctica (Figure 5.9).

Dense, cold water mixes with adjacent waters while flowing around Antarctica. The resulting Antarctic Bottom Water mass flows northward in all three ocean basins. Dense water masses also form in the North Atlantic, near Greenland. Since they are not as dense as the Antarctic waters, they occur at intermediate depths, forming the **North Atlantic Deep Water** (Figure 5.10). These waters flow southward in the Atlantic.

Antarctic Intermediate Water, another water mass of intermediate den-sity, forms near Antarctica. Denser than the surface waters, it flows northward at intermediate depths where the slightly denser North Atlantic Deep Water and **Antarctic Bottom Water** occupy positions below it.

Several other **water masses**, also identified by characteristic temperatures and salinities, form in oceans or marginal seas. For instance, a thin layer of warm saline water (Table 5.1) flows out of the **Mediterranean Sea** through the Strait of Gibraltar and a similar water mass forms in the **Red Sea**. Both form thin layers of warm, salty waters at intermediate depths.

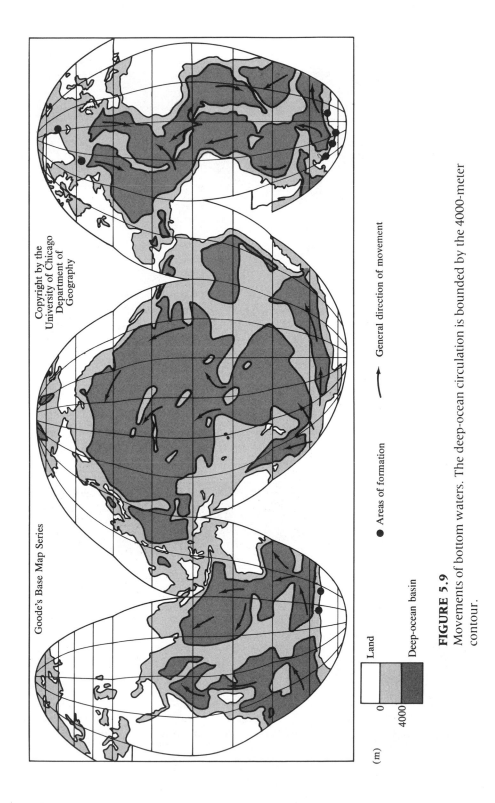

Goode's Base Map Series

Copyright by the
University of Chicago
Department of
Geography

● Areas of formation

→ General direction of movement

(m)

0

4000

☐ Land

▨ Deep-ocean basin

FIGURE 5.9

Movements of bottom waters. The deep-ocean circulation is bounded by the 4000-meter contour.

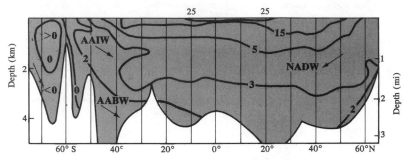

TEMPERATURE (°C)

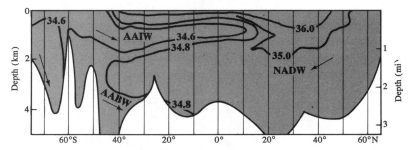

SALINITY (°/oo)

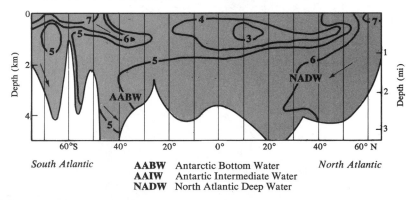

DISSOLVED OXYGEN (ml/liter)

South Atlantic

AABW Antarctic Bottom Water
AAIW Antarctic Intermediate Water
NADW North Atlantic Deep Water

North Atlantic

FIGURE 5.10
Vertical sections showing distributions of temperature, salinity, and dissolved oxygen
in the western Atlantic Ocean (After Wüst). Arrows show general direction of water
movements. [After H.U. Sverdrup, M.W. Johnson, and R.H. Fleming, *The Oceans: Their
Physics, Chemistry and General Biology* (Englewood Cliffs, NJ: Prentice-Hall, Inc., 1942)
p. 748]. Reprinted by permission of Simon and Schuster, Inc.

TABLE 5.1
Some major water masses

	Temperature (°C)	Salinity (‰)
Antarctic Bottom Water	−0.4	34.66
North Atlantic Deep Water	3 to 4	34.9 to 35.0
North Atlantic Central Water	4 to 17	35.1 to 36.2
Mediterranean Water	6 to 10	35.3 to 36.4
North Polar Water	−1 to +10	34.9
South Atlantic Central Water	5 to 16	34.3 to 35.6
Sub-Antarctic Water	3 to 9	33.8 to 34.6
Antarctic Circumpolar Water	0.5 to 2.5	34.7 to 34.8
Antarctic Intermediate Water	3 to 5	34.1 to 34.6
Indian Equatorial Water	4 to 16	34.8 to 35.2
Indian Central Water	6 to 15	34.5 to 35.4
Red Sea Water	9	35.5
Pacific Subarctic Water	2 to 10	33.5 to 34.4
Western North Pacific Water	7 to 16	34.1 to 34.6
Pacific Equatorial Water	6 to 16	34.5 to 35.2
Eastern South Pacific Water	9 to 16	34.3 to 35.1

SOURCE: Albert Defant, *Physical Oceanography*, vol. 1 (New York: Pergamon Press, 1961).

5.8 DEEP-OCEAN CIRCULATION

Below the pycnocline, waters move through the ocean basins in sluggish currents. This deep circulation is almost completely isolated from surface currents by the pycnocline and is driven primarily by differences in seawater density, which are mainly controlled by variations in temperature and salinity. Hence the deep-ocean circulation is called the *thermohaline* (thermo = heat; haline = salt) *circulation*.

Little is known about these slow, deep currents since they are difficult to measure directly (Figure 5.11). Submerged floats can be tracked over long distances using sound signals (Figure 5.12) to study movements of individual water parcels, but the movements are so slow that it is difficult to obtain enough data to map the sluggish currents.

In the deep ocean there is no counterpart to the information obtained from mariner's observations of surface currents. Instead, observed distributions of temperature, salinity, and dissolved oxygen provide most of our information. Heat, salt, and dissolved gases move across boundaries between adjacent layers by mixing and diffusion.

Subsurface water masses move horizontally, primarily in thin layers through the deep ocean. Since each layer is thin, it gradually loses its identity by slowly mixing with adjacent waters. The process apparently takes decades

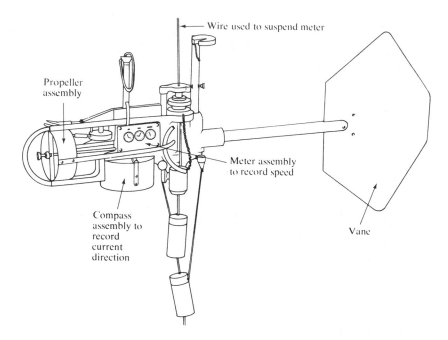

FIGURE 5.11
Current meters measure current directions and speeds, much as anemometers measure wind speed and direction.

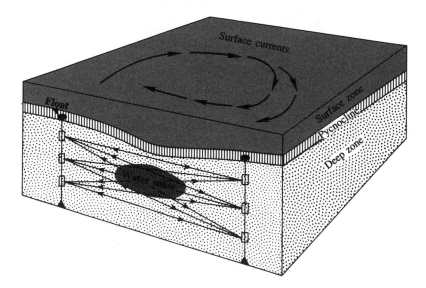

FIGURE 5.12
Acoustic tomography uses multiple sound paths from many sound transmitters and receivers to determine locations and movements of subsurface water masses.

or centuries. The characteristic temperatures and salinities can be used to trace the movements of the water masses.

Deep-ocean currents move generally north-south (Figure 5.9). Like surface currents, deep-ocean or **bottom currents** are strongest on the western side of ocean basins. In contrast to surface currents, bottom currents cross the equator in the Atlantic and Pacific oceans. The subsurface circulation in the Atlantic connects the northern and southern polar regions.

Flows of bottom waters are also influenced by seafloor topography. Gaps in the Mid-Atlantic Ridge channel movements of bottom waters through the deep basins of the western Atlantic into the basins of the eastern Atlantic. Conversely, a ridge separating the Arctic Sea from the Atlantic Ocean prevents flows into the Atlantic of dense water masses formed in the Arctic Sea. Only water masses formed near Greenland (south of the ridge) can flow into the deep Atlantic basins.

Subsurface waters eventually return to the sea surface. This return flow is accomplished by slow, upward movements of deep waters. Much of the return flow to the surface appears to be concentrated in upwelling along the equator and in coastal areas. An unknown amount flows up through the pycnocline.

5.9   MIXING PROCESSES

Mixing of ocean waters results from their continual movement. In the ocean, water movements occur on all scales: Water molecules are constantly moving and are in turn carried by eddies; eddies are transported by currents; and such swirling movements transfer heat, salt, and other properties.

If a water parcel is small, molecular motions cause the mixing. If the parcel is somewhat larger, eddies cause the mixing. Finally, if a water mass is large, currents cause the mixing. *In general, mixing results from water motions the same size as or smaller than the water parcel involved. Water movements on a scale larger than that of the water parcel transport water rather than mix it.*

Energy for mixing comes from winds, currents, and tides. Vigorous mixing is most likely at the sea surface and near the seafloor. At the sea surface, a nearly isothermal (or isohaline) surface zone, often tens of meters thick, is formed by wind-induced mixing. Density changes resulting from surface cooling cause mixing through the sinking of denser waters. This is especially important in polar regions.

Flow near the ocean bottom also causes mixing. Such mixing is especially noticeable when strong currents flow across a sill or ridge. In shallow waters, bottom-associated mixing is especially important, and there is often a zone of well-mixed waters near the ocean floor.

Near coasts, strong tides and currents supply energy for mixing, as do breakers in the surf zone. Fresh water discharged by rivers mixes with seawater in estuaries and near continental margins.

Below the pycnocline in the open ocean, there is relatively little energy available to cause mixing. In a stable situation, the water below will be denser than any given water parcel; the water above will be less dense. In this situation, a water parcel will spread laterally, along surfaces of constant density, as a thin layer interleaved between other layers.

Mixing occurs at the interfaces between layers. Some occurs as a result of different directions and speeds of the movements of the different layers. Mixing also occurs as a result of diffusion of molecules between adjacent layers.

QUESTIONS

1. List some major eastern and western boundary currents.
2. Explain why upwelling is common along the equator.
3. Describe the Coriolis effect.
4. What is a geostrophic current? Are all major ocean currents geostrophic?
5. Describe the general deep-ocean circulation patterns.
6. What evidence supports the statement that the oceanic and atmospheric circulations are closely linked?
7. Draw a simple current gyre. Label the eastern and western boundary currents as well as the east-west currents for the North Atlantic and the South Pacific oceans.
8. Where is the densest bottom water in the ocean formed? How?

SUPPLEMENTARY READING

Books

Barry, R. G., and Chorley, R. J. *Atmosphere, Weather, and Climate.* New York: Holt, Rinehart and Winston, 1970. Introduction to climatology.

Stowe, Keith S. *Ocean Science.* New York: John Wiley & Sons, 1979. Elementary, good discussion of physical processes.

Articles

Gordon, A. L., and Comiso, J. C. "Polynas in the Southern Ocean." *Scientific American* 258(6):90–97.

McDonald, J. E. "The Coriolis Effect." *Scientific American* 186(5):72-78.

Stewart, R. W. "The Atmosphere and the Ocean." *Scientific American* 221(3):76–105.

KEY TERMS AND CONCEPTS

Major surface current systems	Coriolis effect	Ocean topography
Gyres	Ekman spiral	Deep ocean circulation
Eastern boundary currents	Geostrophic current	Acoustic tomography
	Upwelling	Water masses
Western boundary currents	Sinking	Mixing processes
	Rings	

6
Waves and Tides

The ocean surface is rarely still. Disturbances ranging from gentle breezes at the surface to earthquakes many kilometers beneath the ocean bottom can generate waves. *Winds, earthquakes, and the attractions of the sun and moon are the waves' three most important generators.*

Winds cause waves that range from ripples less than 1 cm high to giant, storm-generated waves more than 30 m (100 ft) high. Tides also behave like waves but are so large that their wavelike characteristics are not easily seen. Seismic sea waves, caused by earthquakes, cause catastrophic damage and loss of life, especially in lands bordering the Pacific Ocean. In this chapter, we first examine ideal waves and then their behavior in the ocean.

6.1   IDEAL PROGRESSIVE WAVES

Let us begin by studying waves in a series called a wave train (Figure 6.1). **Progressive waves** passing a fixed point show a regular succession of **crests**—the highest parts of waves—and **troughs**—the lowest parts. **Wave height** (H) is the vertical distance from a crest to the next trough. Successive crests (or troughs) are separated by one **wave length** (L). The time required for successive crests (or troughs) to pass the fixed point is the **wave period** (T), commonly expressed in seconds. Wave period is often used to classify waves.

Wave speed (V) is calculated by V = L/T. This simple formula says that wave speed, wave length, and wave period are all directly related; knowing

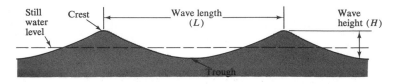

FIGURE 6.1
Simple wave and its parts.

any two factors, we can calculate the third. Wave height is unrelated to the other three factors and must be observed. **Wave steepness**—expressed as H/L—is the ratio of wave height to wave length.

A cork floating on the water surface moves forward on each wave crest and backward in the wave trough. After the passage of each complete wave, the cork returns to its initial position. Only the wave form moves; *in deep water there is little net water movement associated with the wave's passage.*

Movements of small markers floating at various depths in a tank show that water moves in nearly circular vertical orbits as waves pass (Figure 6.2). Orbital diameter equals wave height. Beneath wave crests, particle motion is momentarily in the direction of wave motion. In wave troughs, particle motion reverses. Away from the surface, speed or **orbital motion** decreases and orbits become smaller. At a depth of half a wave length (L/2), orbital motion nearly vanishes.

Water particles move slightly faster in wave crests than in wave troughs so there is a slight net movement of water in the direction of wave travel. Water movement, however, is much slower than wave speed, and for most considerations of wave processes we ignore these small displacements.

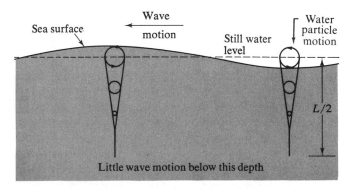

FIGURE 6.2
Wave profile and water particle motions caused by a wave in deep water. Note the diminishing size of the orbits with increasing distance below the surface.

6.2 SHALLOW-WATER WAVES

Where water depths are less than half a wave length, the bottom interferes with wave-induced water motions. Water particles near the bottom cannot move vertically, only horizontally (Figure 6.3). Further from the bottom, water particles move in flattened elliptical orbits that are flatter near the bottom and more circular near the surface.

Waves are unaffected by the bottom where water depths exceed one half the wave length. Therefore in the deep ocean, wave speed is determined by the wave length and period. Longer waves travel faster than the shorter ones. Long waves from a distant storm arrive first, followed by shorter waves.

In shallow water, water depths control wave speed. Where the water depth is less than 1/20 of the wave length, wave speed (V) is controlled by average water depth (d) and can be calculated by $V = 3.1 \sqrt{d}$. As waves move from deep water into shallow water, wave speed and wave length change, but the wave period does not. The direction of wave advance also changes when the wave meets a shallow or irregular bottom.

As waves approach a beach and the water shallows (becomes less than L/2 in depth), the bottom influences orbital water-particle motions. Although the wave period remains unchanged, the wave length is shortened. As a result, wave height increases and wave crests become more peaked. The wave steepness (H/L) increases until it reaches a critical value, about 1/7. At this point, the wave crest peaks sharply, becomes unstable, and breaks (Figure 6.4). Waves usually break when water depths are 1.3 times the wave height.

Energy from a breaking wave may cause new, smaller waves to form. These waves also break when they reach shallower water. Thus the surf zone may have several sets of breakers, depending on wave conditions and near-shore bottom configuration.

When waves break, their energy is expended through turbulence and by water washing up on the beach. In these final stages, wave energy is changed

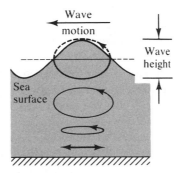

FIGURE 6.3
Motions of water particles caused by shallow-water waves.

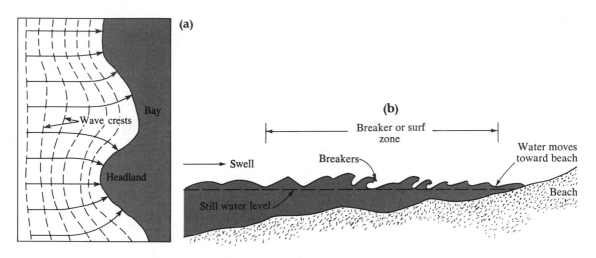

FIGURE 6.4
a. Waves approaching a shore are diffracted by varying water depths. In the deeper waters of a bay, wave crests are diffracted to spread their energy over a large area. Wave energy is concentrated on the headlands, causing erosion. b. Swell peaks upon entering shallow waters. At a depth of 1.3 times the wave height, the wave breaks. It then reforms and again breaks. Finally the water itself moves up onto the beach. [After U.S. Army Waterways Experiment Station, *Shore Protection Manual* 4th edition, 1984]

to heat energy. If this heat were not thoroughly mixed with large volumes of seawater, water temperatures in the surf zone would rise appreciably.

6.3   SEISMIC SEA WAVES (TSUNAMIS)

Large waves are generated by sudden movements of the ocean bottom, usually caused by earthquakes or explosions. These **seismic sea waves,** or **tsunamis** (pronounced *soo-na '-mees*), have wave lengths up to 200 km (125 mi), periods of 10 to 20 min, and wave heights in the open ocean up to 0.5 m (1.5 ft). Although sometimes called tidal waves, they are unrelated to the tides, which we discuss later in this chapter.

On the open sea, seismic sea waves are small and pass unnoticed by ships. Sea level rapidly changes, however, when the waves encounter certain types of bottom topography in shallow water. For example, a large valley can focus a seismic sea wave, causing an enormous breaker. Large loss of life and extensive property damage have resulted from tsunamis. Japan, Hawaii, and Alaska are especially susceptible to these catastrophic waves. In the past 150 years, the Hawaiian Islands have, on the average, experienced a seismic sea wave every four years.

6.4   WIND WAVES

Winds blowing across a still water surface form small wavelets or ripples usually less than 1 cm high with rounded crests and V-shaped troughs. Because of the small size of ripples, **surface tension**—resulting from the mutual attractions of water molecules—influences their shape. Ripples, also called **capillary waves**, move with the wind and last only a short time, but they provide much of the wind's grip on the water surface.

As wind speed increases, small **gravity waves** form and travel with the wind. The size of the waves formed by the wind depends on its speed, the length of time it blows in one direction, and the distance (called the **fetch**) it has blown across the water (Figure 6.5). In short, *wave sizes depend on the amount of energy imparted to the water surface by winds.*

In a storm, a complicated mix of superimposed waves and ripples, known as a **sea**, develops. Waves continue to grow until they are as large as a wind of that speed can generate. After the winds die, the waves continue moving away from the generating area. After leaving the generating area, the waves change, becoming more regular. Long, regular waves outside the generating area are known as **swell** (Figure 6.5).

Wind waves are classified according to their period. Ripples have periods of a fraction of a second. Wind waves in fully developed seas have periods up to 15 seconds; swells have periods of 5 to 16 seconds. Unlike wind-generated currents, wind waves are not affected by the Coriolis effect, because little water movement is associated with their movements.

Waves move in the same direction as the winds that generate them. A wind from a different direction will destroy an earlier wave pattern and generate a new one in its place. Little energy is lost by waves crossing the deep ocean;

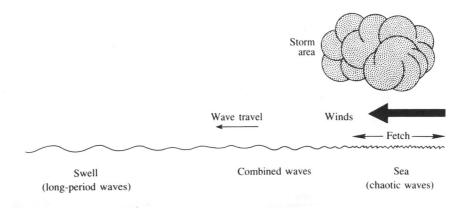

FIGURE 6.5
Schematic diagram showing wind-wave development in a storm and changes in wave profiles as the sea evolves into swell while moving out of the wave-generating area.

waves continue until they meet an obstacle where their energy is dissipated on a cliff, beach, or breakwater.

6.5   STATIONARY WAVES

In addition to progressive waves, there are **stationary waves**—where the wave form does not move through the water—which occur throughout the ocean. These waves are also known as **seiches** (pronounced *say-shes*) and are easily generated in a round-bottomed coffee cup by tilting the filled cup and setting it down on a surface. Viewed from the side, the water surface tilts toward one side and then toward the other. This oscillation of the water surface is an example of a standing wave. When you spill coffee from a cup you are carrying, the culprit is usually a standing wave.

During each oscillation, the water surface remains at the same level at certain locations called **nodes** (Figure 6.6). The stationary waves generated in a small cup or dish usually have a single nodal line where the water level does not change. It is also possible to have several nodal lines or nodal points about which the water surface tilts.

At the **antinodes**, or **crests**, vertical movement of the water surface is greatest. Many antinodes or crests can occur in a container or body of water, but they always occur at the ends of the container or basin. By placing chips or dye in the water, we can observe water motions generated by stationary waves. There is no water movement when the water surface is tilted most. When the water surface is horizontal, its equilibrium position, it moves most rapidly . The largest horizontal water movements occur below the nodal line; beneath the crests, wave movements are entirely vertical.

Stationary waves may be generated in an enclosed body of water by a sudden disturbance, such as a storm or sudden change in atmospheric pressure. Once set in motion, a body of water will oscillate with a period determined by the depth and length of the basin. Eventually the oscillation dies down due to the friction of the water moving across the bottom. Lake Erie has a periodic change of water level of 8 cm with a period of 14.3 hours; that is, every 14.3 hours the lake surface returns to its original position. In Lake Michigan the change in level is 7 cm with a characteristic period of 6 hours. Because of the

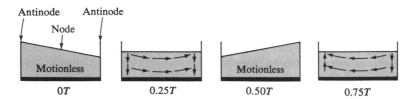

FIGURE 6.6
Water motions in a simple stationary wave at quarter-period intervals.

Earth's rotation, wave crests in large lakes in the Northern Hemisphere move around the basins in a clockwise direction.

6.6 TIDES

Among ocean phenomena, **tides**—*the periodic rise and fall of the sea surface*—are easiest to observe. A firmly anchored pole, marked in meters (or feet), can be used to measure the relative height of the tide at frequent intervals, and tide records can be made from these data. A simple tide gauge consists of a float connected to a pencil, which draws a **tidal curve** (a record of sea level over several days) on a paper-covered, clock-driven cylinder. Modern tide gauges measure the pressure at the bottom of a water column and record the data digitally for analysis by computers. Some even transmit tidal readings by satellite to a central facility, where they are analyzed immediately to detect tsunamis.

Although well known since antiquity, the astronomic origin of the tides was first explained by Sir **Isaac Newton** (1642–1727). His law of gravitational attraction states that the attraction between two bodies is directly proportional to the product of their masses and inversely proportional to the square of the distance between them. In other words, the attraction between bodies is greater the larger the bodies and is smaller as the distance between them increases.

To understand tides, we must consider the gravitational effects of the sun and moon. These two bodies are most important to ocean tides because of the sun's large mass and the moon's nearness. Newton used his theory of gravity to develop an equilibrium model of tides for a water-covered Earth without continents. First we consider the moon's effects alone.

Tides are caused primarily by the gravitational attraction between the Earth, sun, and moon. Tides occur in the solid Earth, in ocean waters, and in the atmosphere, but we will consider only ocean tides. Tides in the ocean are caused by slight differences along the Earth's surface in the gravitational attraction and centrifugal forces between the Earth and moon. (We will ignore the sun for the time being.) The water surface is deformed by these forces, forming an egg-shaped envelope. The Earth rotates within the deformed water envelope, causing the rise and fall of the tides at any location.

On the side of the Earth nearest the moon, ocean water is drawn toward the moon because the distance between the two is slightly less than at the Earth's center. As a result, the water surface is pulled by gravity to form a bulge on the side nearest the moon. On the side opposite the moon, the gravitational attraction is slightly less than at the Earth's center. These centrifugal forces deform the water surface, forming a bulge opposite to the moon. The two tidal bulges are the areas of high tide, and between the two tidal bulges are the troughs or low tides (Figure 6.7).

The moon passes over any location once every 24 hours, 50 minutes (one tidal day). On a water-covered Earth, any point would pass beneath the two tidal crests and the two tidal troughs during each tidal day. If the moon were

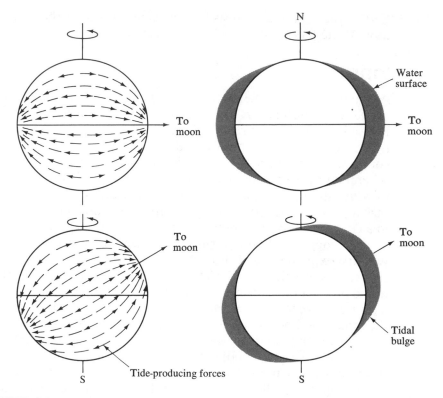

FIGURE 6.7
Tide-producing forces and resulting tidal bulges on a water-covered Earth with the moon in the plane of the Earth's equator (upper) and above the plane of the equator (lower).

in the plane of the Earth's equator, the two high waters at each location would be equal. However, the moon's position (and associated tidal bulges) shifts 28.5° north of the equator to 28.5° south of the equator. This changes the relative heights of high and low waters at any point. When the moon is not in the plane of the equator, there will be one high and one low tide each day.

The sun also has an effect on tides but it is smaller than the effects of the moon. Interactions between the effects of the sun and moon account for some of the complexity in predicting tidal behavior. At certain times during the moon's travel around the Earth, the sun and moon act together (Figure 6.8). At these times, the bulges, or crests, are highest and water levels in the tidal troughs between them are lowest. These are the **spring tides,** when the daily tidal range—vertical distance between high and low tides—is largest. When the moon is near its first and third quarters, solar and lunar tides coincide, and the daily tidal range is the least. These are called **neap tides.**

This theory, developed by Newton, explains the relative effects of the sun and the moon on ocean tides. It also explains why there are two high and two

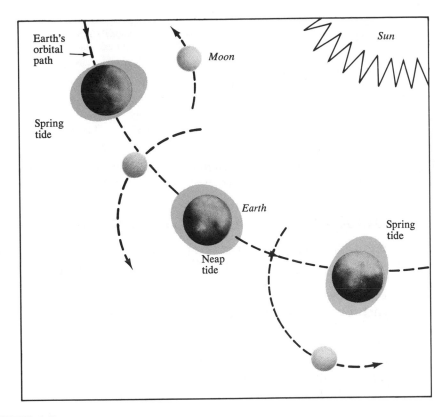

FIGURE 6.8
Relationships of the Earth, moon, and sun during spring and neap tides. Note that the moon and sun are on a line with Earth during times of spring tides but act at right angles to each other during neap tides.

low waters each day in many locations. It fails to predict, however, two important aspects of the tides: their height and the time of high and low tide relative to the moon's passage overhead.

Consider the timing of high water. Our simple equilibrium model predicts that high water will occur when the moon is highest in the sky above our position or direction below our position on the other side of the Earth. This would require that each tidal bulge travel about 1650 km/hr (1050 mi/hr) to keep pace with the moon.

Tides are very long waves. The two tidal bulges, or crests, occur on opposite sides of the Earth so the tides have wave lengths of nearly 22,000 km. Because the ocean is slightly less than 4 km deep, *tides behave like shallow-water waves.* For a tidal wave to travel fast enough to keep up with the moon, the ocean would need to be at least 22 km deep. Consequently tidal crests are displaced from the equilibrium position by frictional drag on the ocean bottom and by the Earth's rotation (Figure 6.9).

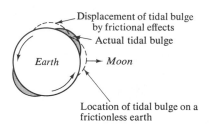

FIGURE 6.9
Position of the tidal bulge is determined by the equilibrium between the moon's gravitational attraction and frictional drag on a rotating Earth (viewed from the North Pole).

6.7   OCEAN RESPONSE TO TIDE-GENERATING FORCES

If the Earth were smooth and water-covered, or ocean basins simple and regular in shape, ocean tides would behave in simple, easily predictable ways, following Newton's laws. As we know, the Earth is only partly water-covered and is far from smooth. Ocean basins have a variety of sizes and shapes. Therefore each basin responds differently to the tide-generating forces.

Tidal height and timing vary greatly between different basins and even at different places within the same basin. Tides in ocean basins are much like a set of stringed instruments designed by a mad artist. The strings on all the instruments are tuned to the tide-generating forces, but the different sizes and shapes of instruments produce a variety of sounds.

Tides in ocean basins vary according to a few main rules:

1. If the characteristic period of the standing wave in a basin is short relative to the period of the tide-generating forces, there is ample time for the water level to be displaced in step with the tide-generating forces. Such a basin has an equilibrium tide.
2. If the characteristic period of the standing wave is very long relative to the period of the tide-generating forces, there is not enough time for the water level to keep step with the tide-generating forces. Therefore, the tides are small and reversed. In other words, low tide occurs when we would have predicted high tide (and vice versa), based on equilibrium tide theory.
3. When the characteristic period of a standing wave in a basin is nearly the same as the tide-generating forces, high and low tide occur about when we would have predicted, but the height of the tide is much greater than expected. The closer the correspondence between the two, the larger the tidal range.

The tide observed at any locality is a combination of standing and progressive waves. In some basins, one or the other predominates. In the Red Sea and Long Island Sound, where the basins are much longer than they are wide, the tide acts like a standing wave with nearly simultaneous rise or fall from one end to the other. High tide or low tide occurs everywhere at nearly the same time. Timing of tidal phenomena is controlled by the tidal wave coming in from the open ocean.

In Chesapeake Bay or Puget Sound, the tide acts like a progressive wave. High water occurs first at the mouth of the system and then advances inland like the crest of a progressive wave. This takes many hours and there may be several crests (or high-water areas) in the system at any time. Each crest is separated by an area of low water, corresponding to the trough of a simple wave.

A similar situation occurs in each ocean basin. Tides move through the ocean like progressive waves. Timing of the movement is controlled by considerations that we have discussed. But the path of the wave and its behavior in any part of the basin are strongly controlled by local effects.

Tidal prediction requires a long period of observations (many months to many years) at a particular point. From these observations, it is possible to determine how that part of an ocean basin responds to the various tide-generating forces and how this response is affected by the complexities introduced by basin shape, waves coming from adjacent ocean basins, friction of the water in the basin, and the Earth's rotation. Tidal tables are the most familiar example of successful predictions of oceanic phenomena.

6.8 TYPES OF TIDES

Despite the complexity of ocean basin responses to tide-generating forces, there are only three types of tides. **Diurnal tides**—one high water and one low water per tidal day—are the simplest (Figure 6.10). This type of tide is common in parts of the northern Gulf of Mexico and Southeast Asia. **Semidiurnal tides**—two high waters and two low waters per tidal day—are common on the Atlantic coasts of the United States and Europe. Note that successive high-water and low-water levels are approximately equal, as predicted by Newton's equilibrium theory. Along the Pacific coast of the United States **mixed tides** are the most common. Successive high-water and low-water stands differ appreciably (Figure 6.11)—there are higher high water and lower high water, as well as higher low water and lower low water. The complexity of tidal phenomena is seen in tidal curves (Figure 6.12) for one month at New York (semidiurnal tide) and Seattle (mixed tide).

6.9 TIDAL CURRENTS

Like other waves, tides cause horizontal water movements called **tidal currents**. In the open ocean away from obstructions, tidal currents constantly change direction (Figure 6.13a) and are known as **rotary currents**. The currents repeat this cycle once each tidal period.

FIGURE 6.10
Diurnal and semidiurnal types of tides.

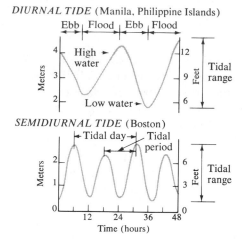

DIURNAL TIDE (Manila, Philippine Islands)

SEMIDIURNAL TIDE (Boston)

FIGURE 6.11
Examples of the mixed type of tide.
F = flood current; E = ebb current.

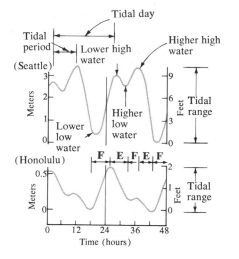

FIGURE 6.12
Tidal variations during one month.

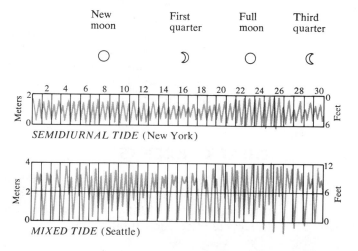

New moon	First quarter	Full moon	Third quarter

SEMIDIURNAL TIDE (New York)

MIXED TIDE (Seattle)

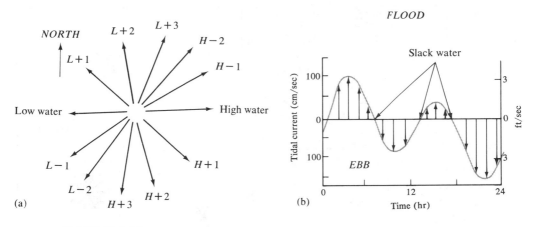

FIGURE 6.13

a. Hourly direction and speed of rotary currents during one tidal period at Nantucket Shoals off the New England coast. Times are shown in hours before and after low and high tide. b. Reversing tidal currents at Admiralty Inlet, Puget Sound, Washington. [After U.S. Naval Oceanographic Office. *American Practical Navigator.* H. O. Publication 9 (Washington, D.C., 1958), p. 712]

Near the shore or in rivers or harbors, the coast obstructs tidal currents (Figure 6.14), preventing the rotary motions observed in open-ocean tidal currents. Here we observe the familiar **reversing tidal currents**. Currents flow in one direction during part of the tidal cycle and reverse their flow during the remainder (Figure 6.13b).

When the water level is rising in a harbor (Figures 6.10 and 6.11), water flows toward the land. There, tidal currents flowing shoreward (upstream in coastal rivers or estuaries) are called **flood currents**. As the tide goes out and sea level falls, water flows seaward. Seaward currents (downstream in coastal rivers or estuaries) are called **ebb currents**. Periods of **slack water**—little or no current—separate the ebb and flood (Figure 6.13b).

Prediction of tidal currents—like that of the tides—is largely based on experience rather than theory. Tidal currents are affected by winds and river flow. In addition, the time and speed of maximum flood or ebb current may vary widely within a single bay or harbor. Tidal current tables contain predictions of these currents, usually based on long series of observations.

6.10   INTERNAL WAVES

Internal waves occur (Figure 6.15) at density discontinuities where layers of different densities meet. They are detected in various ways. Among the easiest to observe are slicks—elongate areas of calm water. These are convergences caused by internal waves where debris and surface-active substances such as oils and greases collect.

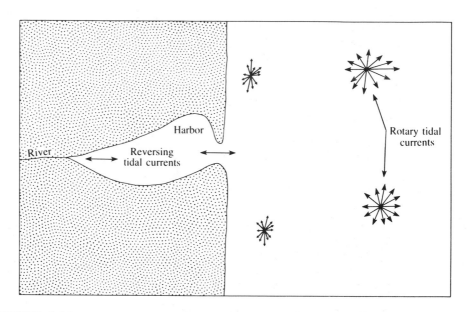

FIGURE 6.14
Schematic representation showing the relationships between offshore rotary tidal currents and reversing tidal currents in bays and rivers. Note that near the coast, tidal currents are transitional between the two, with the strongest currents paralleling the coast.

Temperature or salinity measurements in the pycnocline often reveal internal waves. There, the internal waves cause periodic shallowing or deepening of water having a characteristic temperature or salinity value. Tidal phenomena cause some internal waves in the ocean; other factors doubtlessly also contribute. Because of the difficulty in observing internal waves in the deep ocean, they are poorly understood.

6.11   RIP CURRENTS AND STORM SURGES

Other relatively small net water movements are caused by waves. At the coast, wave-associated water movements toward land cause water to build up near beaches. This causes return flows of water away from the beach. These flows, known as **rip currents** (Figure 6.16), consist of three parts: rip current, rip lead, and return flow toward beach. Shoreward, wave-induced water movements feed water toward the beach.

In the surf zone, the water flows nearly parallel to the beach. Eventually, these longshore currents combine to form jet-like streams, each a few tens of meters across, flowing through the surf and returning the water seaward. Rip currents extend a kilometer or more from the beach before the flow becomes too diffuse to be recognizable.

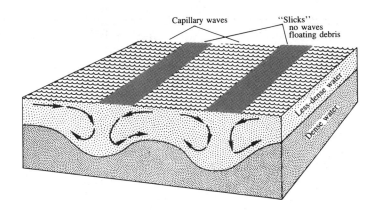

FIGURE 6.15
Internal waves cause water movements that collect floatable materials and surface-active substances (oils, soaplike compounds) above the troughs. These materials inhibit capillary waves, causing slicks—areas of ripple-free water—to mark the presence of the internal waves.

On a beach, rip currents can often be spotted by the absence of surf. Rip currents usually occur in channels deeper than the adjacent ocean floor, and because of the greater depths in these channels, waters in rip currents are usually darker.

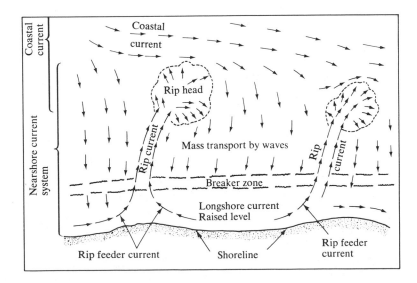

FIGURE 6.16
Rip currents and longshore currents caused by wave action. Currents are strong within the rip currents but diminish outside the breaker zone. [After F.P. Shepard. *The Earth Beneath the Sea*. rev ed. (Baltimore: The Johns Hopkins Press, 1967), p. 18]

Storm surges are changes in sea level caused by storms. Around islands in the open ocean, such storms cause only small changes in sea level (usually less than a meter) because the absence of boundaries in the open ocean permits nearly unrestricted water flow. Storm surges are essentially the opposite of upwelling.

On low-lying coasts, however, storm surges can raise sea level several meters and cause catastrophic flooding. Storm surges killed 250,000 people in 1864 and 1876 around the Bay of Bengal in the northern Indian Ocean. In 1953 a severe North Sea storm caused sea level to rise about 4 m. Storm waves overtopped and breached the dikes protecting the Dutch coast. About 25,000 km² (9500 sq mi) were flooded, 2000 people were killed, and 600,000 had to be evacuated.

QUESTIONS

1. Why are tsunamis most common in the Pacific Ocean?
2. What three processes generate waves in the ocean?
3. During an eclipse of the moon, would we have spring or neap tides? Why?
4. List and briefly discuss the two major types of wind waves. Indicate the dominant force that determines the wave's shape and speed.
5. Contrast an ideal stationary wave to an ideal progressive wave.
6. Explain how the equilibrium theory of the tide explains the semidiurnal tides.
7. List some of the wave characteristics you could use to distinguish between sea and swell.

SUPPLEMENTARY READING

Books

Bascom, W. *Waves and Beaches: The Dynamics of the Ocean Surface,* rev. ed. Garden City, NY: Anchor Books, Doubleday and Company, Inc., 1980. Elementary.

Clancy, E. P. *The Tides: Pulse of the Earth.* Garden City, NY: Doubleday, 1969. Elementary.

Defant, A. *Ebb and Flow: The Tides of the Earth, Air and Water.* Ann Arbor: The University of Michigan Press, 1958. Intermediate in difficulty.

Marmer, H. A. *The Tide.* New York: Appleton-Century-Crofts, 1926. A standard reference.

Russell, R. C. H. and MacMillan, D. H. *Waves and Tides.* New York: Philosophical Library, Inc., 1953.

Tricker, R. A. R. *Bores, Breakers, Waves and Wakes.* New York: American Elsevier Publishing Company, Inc., 1965.

Articles

Bascom, W., ''Ocean Waves.'' *Scientific American* 201(2):89–97.

Goldreich, P. ''Tides and the Earth-moon System.'' *Scientific American* 226(4):42–57.

Truby, J. D. ''Krakatoa—The Killer Wave.'' *Sea Frontiers* 17(3):130–139.

KEY TERMS AND CONCEPTS

Progressive wave

Wave train

Tsunami

Capillary waves

Stationary waves

Wind waves

Fetch

Sea

Swell

Types of tides:
  Diurnal,
  Semidiurnal,
  Mixed

Tidal currents:
  Flood currents;
  Ebb currents

Internal waves

Beach processes

Storm surges

7
Coasts and the Coastal Ocean

Coastal oceans and adjacent seas are only a small part of the ocean—12.5% of the Earth's surface and only 4% of the ocean volume. As land dwellers, we are most likely to see and to use the coastal ocean. These highly productive waters provide food and recreation, and their ocean margins receive wastes from farms, cities, and industries. Coastal waters dilute the waste, and tidal currents disperse it along the coasts. In this chapter we will examine coastal ocean processes and how they affect the environment.

7.1   COASTS

Coasts are where land meets ocean. Since much of the world's population lives within a few tens of kilometers of the ocean, coasts are very familiar to us. Scientists divide coastlines into two types (Figure 7.1). One is formed by marine processes: by wave erosion, sediment deposition, or the effects of marine plants or animals. The other is produced primarily by processes acting on land. Examples are coasts formed by volcanoes and the uplift or sinking of land caused by earthquakes and sea level (Figure 7.2).

We consider first marine processes. Coastlines resulting from sand deposition are common (Figure 7.1). They occur along the Atlantic and Gulf coasts of North America. The sediments come from discharges of rivers or from erosion of headlands and are deposited by currents moving along shores to form barrier islands.

Cape Cod→

Long Island Sound→

Delaware Bay→

Chesapeake Bay→

Cape Fear

ATLANTIC OCEAN

FIGURE 7.1
The U.S. Atlantic coast exhibits complex coastlines formed primarily by terrestrial processes north of Chesapeake Bay and by marine processes to the south. Glaciers cut Long Island Sound and most of Cape Cod. Delaware Bay and Chesapeake Bay were ancient river valleys, flooded when sea level rose to its present level. (Courtesy NASA)

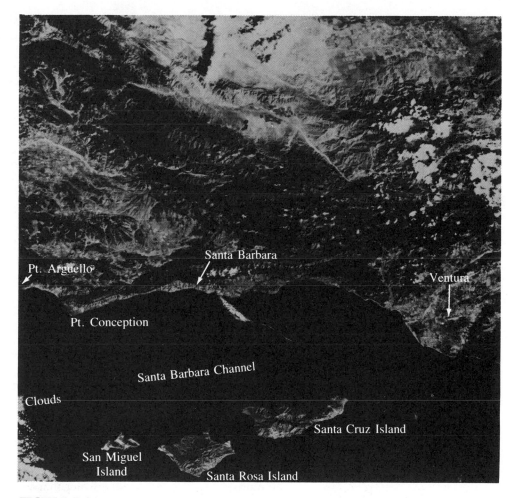

FIGURE 7.2
In Southern California, mountain building results in rocky shorelines with narrow beaches. The continental shelf is as mountainous as the land, with many silled basins, some filled with sediment. Several basins contain major oil fields. (Courtesy NASA)

Marine processes smooth shorelines. In the process, they isolate indentations and form lagoons (elongate, shallow water bodies parallel to the shoreline). They also partially isolate estuaries (arms of the sea where river water and seawater mix).

The rocky shorelines of New England were formed by land processes—continental glaciers that deeply eroded the land and removed the soils. After the glaciers retreated, the sea rose to its present level a few thousand years ago. That is too little time for waves to have eroded the rocky shores or to have formed barrier islands. The rocky shorelines of the U.S. Pacific coast are the result of continued mountain building raising the shoreline. In most areas, the ocean processes have not been able to keep up with the vertical movements of the land.

Another example of shorelines formed by land processes are deltas, the low-lying marshes, wetlands, and swamps formed by sediment deposits, usually at a river mouth. Deltas are normally covered by water-loving, salt-tolerant plants, which trap sediment. The Mississippi Delta was formed by this process. When sediment supplies from the river are reduced or cut off, the sea erodes the low-lying lands. Alteration of the Mississippi Delta during oil and gas exploration and by regulation of river flow to prevent flooding has reduced the sediment supply to the delta, which is now shrinking due to erosion. Where there are adequate sediment supplies, wetlands can keep up with a rising sea level or even build seaward.

Shorelines are affected by the continued rise of sea level. Sea level has been slowly rising for thousands of years since the last retreat of the glaciers. The warming of ocean waters due to the greenhouse effect is also causing sea level to rise more rapidly. This is likely to cause flooding of low-lying coastal areas.

TABLE 7.1
Major features of the coastal and open ocean

Feature	Coastal Ocean	Open Ocean
Depth	To 1 km	4 to 6 kms
Distance	Tens of kms from coast	Hundreds of kms from continental slope
Surface currents	Seasonal; parallel to coastlines	Little change; form nearly circular gyres
Deep currents	Estuarine circulation common	Dominantly north-south; water masses form in high latitudes
Salinity	River discharge dominates; varies seasonally	Precipitation/evaporation dominates; little seasonal variation, only in surface zone
Temperature	Varies seasonally; variations affect bottom	Variation only in surface zone; bottom temp. 2-3 °C, controlled by water mass formation in high latitudes
Sediments	River discharge dominates; most sediment deposited in estuaries and on continental shelf/slope	Atmospheric transport important; most sediment deposited near margins, and under areas of high biological productivity

7.2 CHARACTERISTICS OF THE COASTAL OCEAN

In the open ocean, most processes act over years to centuries. A glass float carried by surface currents takes several years to cross the North Pacific Ocean. Waters take hundreds of years to return to the surface after sinking in polar regions.

The coastal ocean, however, responds to processes acting on it from over a few hours for winds and tides to a few weeks for seasonal changes in temperature and river discharges (Table 7.1). Distances involved in coastal ocean processes are also shorter than those involving the open ocean and coastal waters are often partially isolated from the open ocean. The Sea of Japan's partial isolation from the Pacific Ocean by the Japanese islands is one example. Bays, harbors, and fjords have restricted communication with the sea. Therefore, it is not surprising to find that *near the coast, surface seawater temperatures and salinities can change radically within distances of a few tens or hundreds of kilometers*. Near the coast, tidal ranges are larger and tidal currents are stronger than in the open ocean.

The coast affects the ocean in yet another way. There, wind-induced upwelling results from restricted water flow, as shown in Figure 5.8. In the absence of a coast, only a thin layer of surface water would be affected. Due to the land's obstructing water flows, waters from 100 to 200 m (330 to 650 ft) deep move upward to replace surface waters moved seaward by winds. As we shall see in the next chapter, this upwelling supports the growth of floating plants and animals.

7.3 SALINITY AND TEMPERATURE

Large variations in salinity are common in coastal surface waters. Surface-water salinities are lowest near continents, except in areas of wind-induced upwelling, and are highest near the centers of the open-ocean current gyres. Belts of low surface-salinity tend to parallel coastlines.

Temperature variations in coastal waters are controlled primarily by distance north or south of the equator, as in the open ocean. The presence of coasts has less influence on surface-water temperature than on salinity. Surface temperatures in coastal waters are affected by currents transporting warm or cold water and by wind-induced upwelling of subsurface waters. Cold winds from the land cause substantial cooling during winter.

7.4 COASTAL CURRENTS

Winds and discharges of fresh waters by rivers control coastal currents. Wind-driven coastal currents are highly variable because the winds themselves are so changeable. The currents can respond to changes in winds within a few hours. Storms lasting several days can establish extremely strong currents.

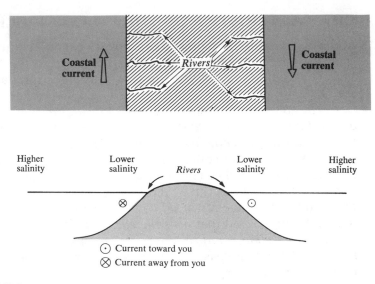

FIGURE 7.3
Sloping sea surface and coastal currents resulting from river discharges into coastal waters in the Northern Hemisphere.

Density variations in coastal waters are dominated by changes in salinity. The salinity structures in coastal waters persist longer than the wind effects. Indeed, one of the primary effects of the wind is to control density distributions in the coastal ocean, which in turn controls geostrophic currents. Remember that the sea surface above low-density waters stands higher than the sea surface above high density waters (Figure 7.3). Low salinity waters along the coast cause a sloping sea surface (ignoring temperature effects on density) highest near the coast and lower near the open sea. Water responds to such a sloping surface as it does in the open ocean. As it runs downhill it is deflected by the Coriolis effect and moves along the sloping surface. This causes currents to flow northward (**Davidson Current**, for example) along the west coast of North America and southward along the east coast (such as the **Labrador Current**).

7.5   ESTUARIES AND ESTUARINE CIRCULATION

Estuaries *are partially isolated basins containing low salinity waters.* They are mixing basins for river discharges and sea waters, and they trap riverborne sediment.

An **estuarine circulation** (long-term nontidal flow) is a mixing system involving a two-way flow of waters. The surface waters (less saline, lower density) move seaward while landward-moving subsurface seawater (higher density) flows in along the bottom of the estuary (Figure 7.4). As the surface

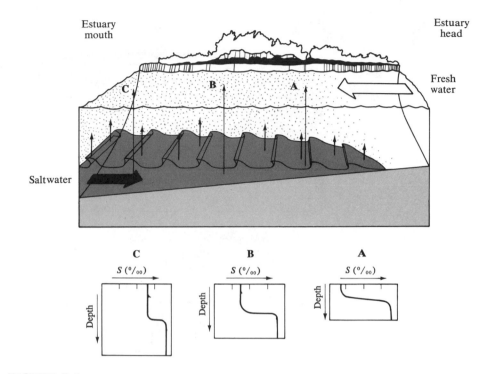

FIGURE 7.4
Vertical water circulation in a simple, salt-wedge estuary. Variation of salinity with depth at three stations is shown in the lower part of the figure. [After D.W. Pritchard, "Estuarine Circulation Patterns," *American Society of Civil Engineers, Proceedings 81* (1955), Separate 717]

waters flow seaward, they entrain seawater from below and mix it with the freshwater discharged by the river. In this way, the surface waters become more saline as they move seaward, and the resulting brackish waters, being less dense than the more saline seawater coming in from the ocean, remain at the surface. This mixing process causes a landward flow along the bottom to replenish the seawater that moved into the surface layers.

Let us compare the amount of seawater flowing in along the estuary bottom to the freshwater contributed by the river. Mixing equal volumes of river water (S = 0 °/oo) and seawater (S = 35 °/oo) results in a salinity of 17.5 °/oo for the mixture. Two volumes of seawater and one volume of river waters result in a mixture with a salinity of 23.3 °/oo. By the time the surface waters leave the estuary they usually have salinities around 30 °/oo, typical of the coastal ocean. In most estuaries, the volume of waters moving in along the bottom is 10 to 20 times greater than the volume of freshwater that rivers discharge.

Where tidal action is large and river discharge small, there is usually more mixing because there is more turbulence. In this type of **moderately stratified**

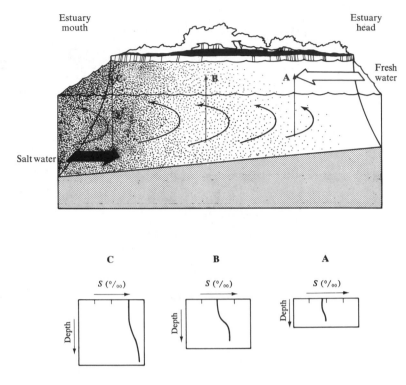

FIGURE 7.5
Variation of salinity with depth at three stations in a moderately stratified estuary. Arrows indicate net (nontidal) water flow. [After D.W. Pritchard, "Estuarine Circulation Patterns," *American Society of Civil Engineers, Proceedings 81* (1955), Separate 717]

estuary (Figure 7.5) the boundary between the less saline surface layer and the more saline subsurface layer is more diffuse. Chesapeake Bay and its tributaries are examples of moderately stratified estuaries, as are the Savannah River estuary in Georgia and Charleston Harbor in South Carolina.

In large estuaries in the Northern Hemisphere, the Coriolis effect deflects the seaward-flowing surface layer to the right. Thus the low salinity waters flow along the right shore (looking seaward) while the high salinity waters occur (still looking seaward) on the left side.

The type of estuarine circulation developed in an embayment is controlled by river discharge, tides, width and depth of the embayment, and presence or absence of a sill at the entrance. As river discharges vary, estuarine circulation patterns also change. For example, during maximum discharge of the Columbia River in Oregon and Washington, the estuary has a **salt-wedge** type of circulation, with the wedge extending only a few kilometers inside the river mouth. At such times, most mixing of fresh and salt waters occurs outside the estuary on the continental shelf. During low-discharge periods salt water is detected near the estuary and less on the continental shelf. The extreme case is the Amazon River where river discharge is so large that salt water is not detectable inside the river mouth. Mixing of salt and fresh water occurs entirely on the continental shelf, outside the estuary. This is an unusual situation since the Amazon is the world's largest river.

Estuarine-like circulation also occurs outside embayments and drowned river valleys. The upward movements of more saline waters into a less saline surface zone is common in large parts of the coastal ocean where surface waters are appreciably diluted by runoff from the land. Large areas of the northern Indian Ocean, the North Pacific Ocean, and the Arctic Sea have estuarine-like circulation patterns. Most continental shelves exhibit an estuarine current pattern except off desert coasts.

7.6 SILLED BASINS

Adjacent land masses restrict water movements in estuaries and inland seas. Many such water bodies are further isolated from the adjacent ocean by a sill, an elevated part of the seafloor partially separating ocean basins. The sill restricts movements of bottom waters, resulting in their partial isolation.

In a basin where surface waters are appreciably less saline than deeper waters, an estuarine-like circulation results. Such a circulation is restricted almost entirely to the waters shallower than the top of the sill (Figure 7.6). The circulation is controlled by the fresh water coming into the basin, by local winds, and by the tides.

Circulation of waters below sill depth is more difficult to predict. This deeper circulation is more sluggish than in the surface zone. Because of the halocline, bottom waters may be little affected by the surface circulation. Water density near sill depth at the basin entrance is the most important factor affecting the deeper circulation. If the water at sill depth outside the basin is appreciably denser than the water in the basin, this denser water flows into the basin and replaces the waters that previously occupied the basin's floor. If the water at sill depth outside the basin is less dense than that inside, the bottom waters are not replaced.

Replacing bottom waters is essential to maintaining life on the basin floor. If bottom waters are not renewed, their dissolved oxygen is consumed by respiration of animals and decay of organic matter that has fallen from the surface layers. When the dissolved oxygen is gone, the only organisms that survive are those that do not depend on dissolved oxygen for their metabolic processes (these are called anaerobes). Some organisms obtain oxygen by breaking down sulfate ions (SO_4^-) in seawater and releasing hydrogen sulfide (H_2S), which is toxic to most organisms. Because of these conditions, the bottom waters in a stagnant basin are nearly devoid of life, except for bacteria and those few organisms that tolerate hydrogen sulfide.

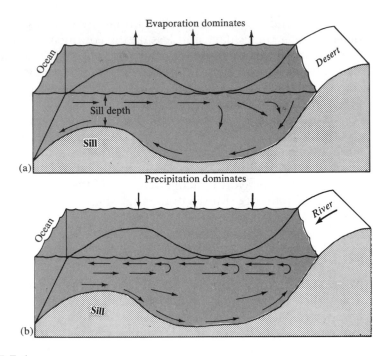

FIGURE 7.6
Water circulation in silled basins where (a) evaporation exceeds precipitation and runoff and (b) runoff and precipitation exceed evaporation.

The **Black Sea** is an example of a stagnant marine basin with an estuarine circulation. Because of the restricted communication with the Mediterranean Sea, Black Sea bottom waters are almost completely isolated. Flow through the Bosporous (the narrow strait connecting the Black Sea with the Mediterranean) would require 2500 years to replace the waters below a depth of 30m. Hydrogen sulfide occurs at all depths below 200 m. Most life in the Black Sea is restricted to surface layers where the waters contain dissolved oxygen.

Water outside silled basins usually contains dissolved oxygen. If the basin's deeper waters are replaced frequently, the dissolved oxygen will not be completely used up. The critical factors are the rate at which the bottom waters are replaced, their initial dissolved oxygen contents, and the rate of oxygen utilization in waters below sill depth.

Some fjords have stagnant or nearly stagnant bottom waters containing hydrogen sulfide. In other silled marine basins, bottom waters are replaced frequently and no stagnation occurs. For example, in Puget Sound—a silled, fjord-like system—the bottom waters are well-oxygenated. Only in the most isolated parts is dissolved oxygen in the bottom waters significantly depleted.

Not all coastal basins have an excess of fresh water. In the Mediterranean and Red seas and the Arabian Gulf, more water is evaporated from the sea surface than is added through river discharge or precipitation. As a result, surface waters become more saline (Figure 7.6). Salinity of surface waters at the head of the Red Sea and Arabian Gulf exceeds 40°/oo, and at the eastern end of the Mediterranean Sea it exceeds 39°/oo.

Where evaporation exceeds precipitation and river runoff, the resulting vertical water circulation is nearly the opposite of the estuarine circulation caused by added fresh water to the surface layers. In evaporating basins, surface waters become denser than those below, and they sink. At the entrance to the basin, there is a net inflow of surface waters to replace the water lost by evaporation. There may also be a subsurface outflow of warm, salty waters.

7.7   SEA ICE

Sea ice is a common feature of the coastal and high-latitude ocean. It forms when sea water is cooled below the freezing point. Microscopic crystals form first and then grow to be hexagonal needles, 1 to 2 cm long. When ice begins to form,

FIGURE 7.7
An icebreaker moves through Antarctic sea ice. (Courtesy NASA)

the sea surface has a dull look and no longer reflects the sky. As freezing continues, a lattice of crystals develops, covering the surface like a blanket of wet snow. Eventually the crystals grow downward and grow together. A thin plastic layer forms, consisting of ice and small enclosed cells containing seawater.

Ice crystals themselves contain no salt but brines occur in small cells between the crystals. Typically 1 kg of newly formed ice consists of about 800 g of ice (salinity 0 °/oo) and 200 g of seawater (salinity 35 °/oo). In other words, the average salt content of newly formed sea ice is about 7 °/oo.

As water temperatures continue to fall, new ice forms on the underside of the ice layer, making it thicker. Brine in the cells also freezes, adding ice to the interior walls of the cells. Remaining brines become more concentrated and if temperatures are low enough, salt eventually crystallizes.

Salt content of newly formed sea ice depends on the temperature of ice formation. At temperatures near the freezing point, ice forms slowly, allowing

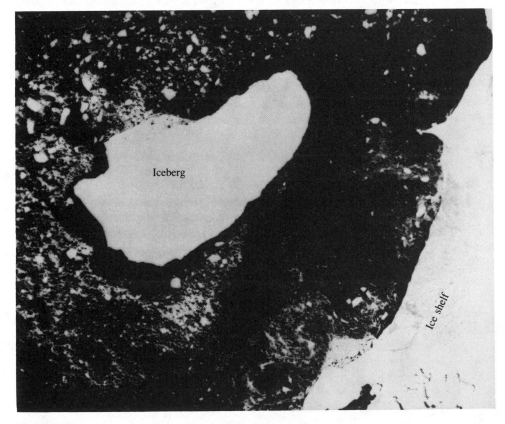

FIGURE 7.8
A gigantic tabular iceberg, about the size of the state of Rhode Island, in Antarctic waters. (Courtesy NASA)

brines to escape. Relatively little seawater remains in the cells and the salt content of the ice is low. At lower temperatures, ice forms more rapidly, trapping seawater so that ice is salty but always less salty than the seawater from which it formed.

As sea ice ages, it excludes salt as the ice hardens. A layer up to 20 cm (about 8 in) thick can form in one winter (Figure 7.7). Over many seasons, the maximum ice thickness is usually 2 to 3.5 m (6 to 11 ft). Thicker pieces of sea ice form when winds pile pieces on top of one another. Very thick areas, called pressure ridges, form where the sea ice pieces deform when pushed together.

7.8   ICEBERGS

Icebergs are fragments of glaciers that flow into the sea. Many icebergs break off the Greenland glaciers each year and move south with the currents into the open Atlantic. The *Titanic* sank in 1912 with heavy loss of life after ramming such an iceberg.

The largest icebergs are formed by pieces breaking off the floating ice shelves around Antarctica. These form enormous tabular icebergs (Figure 7.8) which move with the currents, persisting for several years until they melt.

QUESTIONS

1. Describe a geostrophic coastal current caused by river discharge in the Southern Hemisphere.
2. List some processes that cause mixing of ocean waters. In what part of the ocean is each most important?
3. Describe the circulation in a silled basin where evaporation exceeds precipitation. Give an example.
4. Define estuary. What is the simplest estuarine flow? Describe two different types of estuarine circulation patterns. What causes them to be different?
5. Explain why storm surges occur. Where are they most likely to cause damage to humans?
6. Describe a typical beach profile.

SUPPLEMENTARY READING

Books

Bird, E. C. F. *Coasts*. Cambridge: The M.I.T. Press, 1969. Elementary.

Hay, J. *The Great Beach*. New York: Ballantine Books, 1963. Well-written history of Cape Cod beaches.

King, C. A. M. *Beaches and Coasts*. 2d ed. London: Edward Arnold, 1972. Descriptive.

Shepard, F. P. *Submarine Geology*. 3d ed. New York: Harper & Row, Publishers, 1973. Chapters 6, 7, and 8 deal with the coastal ocean and beaches.

Articles

Bascom, W. "Beaches." *Scientific American* 203(2):80–97.
Carr, A. P. "The Ever-Changing Sea Level." *Sea Frontiers* 20(2):77–83.
Emiliani, C. "The Great Flood." *Sea Frontiers* 22(5):256–270.

KEY TERMS AND CONCEPTS

Scales of coastal ocean
   processes
Salinity and temperature
   variations
El Niño

Coastal currents
Silled basins:
   Excess precipitation;
   Excess evaporation

Sea ice
Rip currents
Estuaries
Estuarine circulation

8
Life in the Ocean

Marine organisms obtain energy and the materials to build their tissues from their surroundings. All marine organisms are parts of **marine ecosystems,** which link plants and animals in the ocean to the environments in which they live.

In this chapter, we examine aspects of oceanic life, various marine organisms, the factors that control their distributions and abundance, and some effects of marine organisms on the composition of seawater.

8.1   LIFE CONDITIONS IN THE OCEAN

Let's begin by comparing living conditions in the ocean to those on land. On land, organisms live primarily on the surface; only birds, insects, and the largest trees extend more than a few meters into the atmosphere. *In contrast, the ocean provides a three-dimensional environment* (Figure 8.1). Living space in the ocean greatly exceeds living space on land and provides many different kinds of life conditions.

Most marine plants and animals live near the top, bottom, or sides of the ocean. Most food in the ocean is produced in well-lighted, near-surface waters (called the photic zone) so life of all kinds is most abundant there. As animals feed, the fecal matter they produce sinks to the bottom. Organisms that grow in near-surface waters also sink toward the bottom after death. Many animals

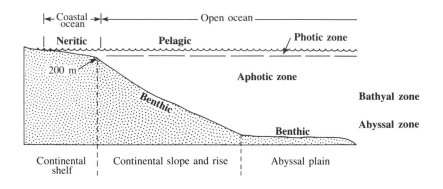

FIGURE 8.1
Major life zones in the ocean.

and bacteria consume these materials as they sink or lie on the ocean floor. Thus food produced in surface waters sustains life throughout the ocean.

Seawater and living tissues have nearly the same density, about 1.03 g/cm³. This provides two important advantages to organisms living in the ocean. First, it permits organisms to float (or sink slowly). Many marine organisms are **pelagic** (floating) or free swimming. To maintain their preferred depth range they must float or swim. The second advantage is the elimination of the need for the heavy skeletons required to support land plants and animals. Since marine organisms do not need heavy skeletons for support, many of them are delicate and jellylike. Jellyfish are familiar examples.

Active swimmers, such as fish, porpoises, or whales, have skeletons similar to those in land-dwelling animals. Plants and animals living on land need skeletons for support and to permit movement since air does not support as seawater does. The viscosity of seawater inhibits rapid motions. Birds can fly through air much faster than fish can swim through water.

Ocean water is not very transparent to light so visibility is limited. Even with bright light and exceptionally clear water, visibility is usually less than 30 m in the ocean. In contrast, we can usually see for many miles through a clear atmosphere. Marine animals that depend on seeing their food must therefore have large, very sensitive eyes. Smelling and sensing vibrations from organisms are also useful means of locating food.

Many organisms emit light from special organs. Some use light to trap unwary smaller organisms. Others use light to frighten predators. When alarmed, some squid release luminous ink into the water to confuse the sight of predators and to permit escape.

Marine organisms range in size from tiny cells too small to be seen by light microscopes to the largest animals on Earth—the blue whale (Figure 8.2). *Most marine organisms are small, averaging about the size of a mosquito.* They are mostly moved by currents, being too small to swim effectively against them.

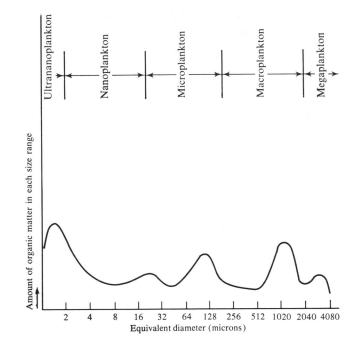

FIGURE 8.2
Relative abundances of different sized marine organisms. [After T.R. Parsons, M. Takahashi, and B. Hargrave, *Biological Oceanographic Processes*, 2d ed. (Oxford: Pergamon Press Ltd., 1977). Used by permission]

Marine organisms have mechanisms that permit them to locate and catch these smaller organisms. They also have filtering devices that permit them to capture food in large volumes of water.

Diversity of life in the ocean varies substantially from place to place. The greatest diversity of marine organisms occurs in tropical waters where there are many different types of organisms but few representatives of each one. Conversely, severe and changeable conditions (subpolar regions, for instance) have relatively few different kinds of organisms but many of each type.

Life is unevenly distributed within the ocean. Large expanses of surface open-ocean waters, far from land, are nearly devoid of life. These ocean areas have exceptionally clear waters, appearing luminous blue in the sun. The Sargasso and the Mediterranean are examples of these colorful, unproductive waters.

Coastal ocean waters (neritic environment) are especially rich in marine life. These relatively shallow waters produce more plant material than do comparable areas of the open ocean. In addition, plant material washes into the ocean from land. Thus, much of the world's fish production comes from coastal ocean waters, especially areas of persistent upwelling.

Marine organisms usually occur in patches. Some patchiness results from processes that concentrate floating organisms into long rows, much the same way that tidal currents sweep floating debris into patches called tide rips. Some patchiness results from plant growth in areas of abundant nutrients, and the abundance of food usually attracts feeding animals.

8.2   MARINE ORGANISMS

Marine organisms are divided into **plankton, benthos, nekton,** and **neuston,** depending on where and how they live. Each group has a different lifestyle and a different role in the ocean.

Ocean plankton includes minute pelagic plants and animals that are weak swimmers and thus moved by ocean currents; they go where currents take them. Marine plankton are further divided into **phytoplankton**—plant plankton—and **zooplankton**—animal plankton.

Phytoplankton are minute single-celled plants. Two major groups— diatoms and dinoflagellates (Figure 8.3 and 8.4)—produce most of the organic matter which supports marine animals. Because they require sunlight for photosynthesis, phytoplankton occur in sunlit, near-surface waters known as

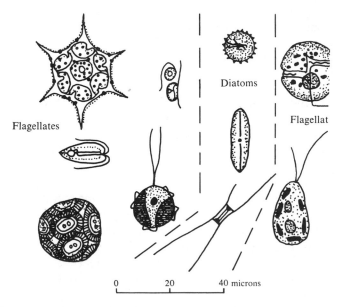

Flagellates

Diatoms

Flagellat

0   20   40 microns

FIGURE 8.3
Common marine plankton: flagellates and diatoms. [After T.R. Parsons, M. Takahashi, and B. Hargrave, *Biological Oceanographic Processes,* 2d ed. (Oxford: Pergamon Press Ltd., 1977). Used by permission]

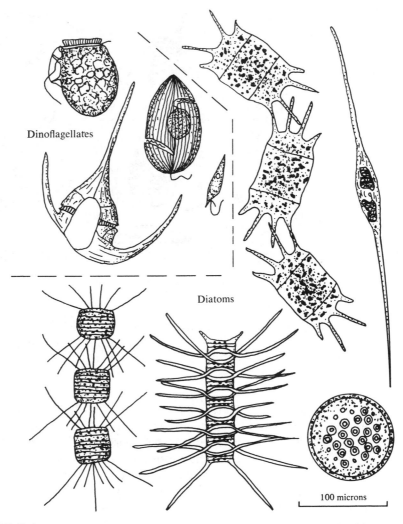

Dinoflagellates

Diatoms

100 microns

FIGURE 8.4
Common marine microphytoplankton: dinoflagellates and diatoms.
[After T.R. Parsons, M. Takahashi, and B. Hargrave, *Biological Oceanographic Processes*, 2d ed. (Oxford: Pergamon Press Ltd., 1977). Used by permission]

the **photic zone**. The thickness of the photic zone is controlled by latitude, season, time of day, and water clarity.

Zooplankton (Figure 8.5) includes two groups. One group, **holoplankton**, spends its entire life as plankton. The second group, **meroplankton**, is planktonic only during certain stages of development, usually the larval stages. Adult stages of the meroplankton live in or upon the ocean bottom.

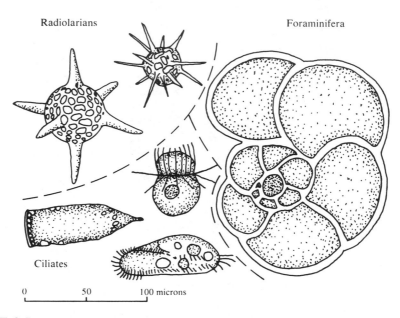

Radiolarians   Foraminifera

Ciliates

0   50   100 microns

FIGURE 8.5
Marine microzooplankton: radiolarians, ciliates, and foraminifera. [After T.R. Parsons, M. Takahashi, and B. Hargrave, *Biological Oceanographic Processes*, 2d ed. (Oxford: Pergamon Press Ltd., 1977). Used by permission]

On continental shelves where waters are shallow, meroplankton are more abundant than in the open ocean's deep waters. The planktonic larvae of benthic species permit organisms to disperse their young widely, allowing them to colonize new areas when conditions are favorable.

Zooplankton include representatives of most major groups of marine animals. Small crustaceans dominate. These include shrimp-like copepods, amphipods, mysids, and euphausiids, as well as larvae of shrimp, crabs, and lobsters (Figure 8.6).

Zooplankton occur at all depths in the ocean. They are most abundant in the surface and near-surface waters. Zooplankton, like phytoplankton, are classified according to the water depths where they live. **Epipelagic** organisms live in near-surface waters, less than 200 m deep; **mesopelagic** forms live between 200 and 700 m; and **bathypelagic** animals live below 700 m. Zooplankton living in deeper waters are larger than near-surface organisms. Many mesopelagic organisms migrate vertically, feeding near the surface at night and returning to deeper waters during the day, presumably to avoid predators.

The **benthos** *are organisms living on or in the ocean bottom* (**benthic environment**). Some organisms (**infauna**) live in the sediments. Others (**epifauna**) live on the bottom or in near-bottom waters. Some organisms bur-

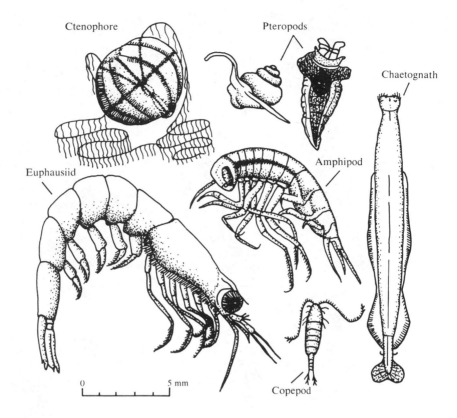

FIGURE 8.6
Larger zooplanktonic organisms: macrozooplankton and megazooplankton. [After T.R. Parsons, M. Takahashi, and B. Hargrave, *Biological Oceanographic Processes,* 2d ed. (Oxford: Pergamon Press Ltd., 1977). Used by permission]

row; others attach themselves to the substrate; and still others crawl about on the bottom or swim freely in near-bottom waters. Members of nearly every major group of marine animals occur in the zoobenthos (Figure 8.7). They include protozoans, various marine worms, mollusks, crustaceans, and starfish. Bacteria are abundant in sediments and also occur living free in the ocean waters.

In relatively shallow water, plants grow on the bottom (**phytobenthos**). Seagrass and marine algae are most abundant in shallow waters. None can grow deeper than about 200m because of the need for sunlight for photosynthesis.

Benthic organisms are far more abundant in shallow waters than on the deep-ocean floor. In near-shore waters, standing crops may be as dense as one kilogram or more of living matter per square meter of bottom. In the deep ocean, living matter amounts to one gram or less per square meter. In the stagnant bottom waters of isolated ocean basins (such as the Black Sea), normal marine

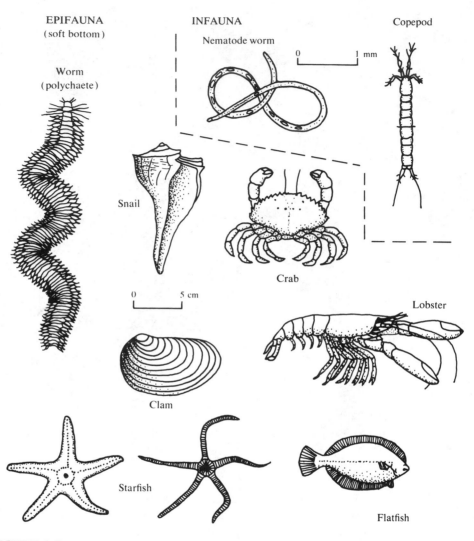

FIGURE 8.7
Some common benthic organisms.

animals cannot survive; only bacteria and a few sulfide tolerant organisms can live.

Nekton *are free-swimming animals that can move independently of currents.* Marine mammals (whales and porpoises), most fish, cephalopod mollusks (especially squids), and some swimming crustaceans are nekton (Figure 8.8). Nekton are most abundant in near-surface waters but occur at all depths in the ocean. Like some of the mesopelagic organisms, many nektonic organisms have daily vertical migrations. They feed on the abundant food in near-surface waters

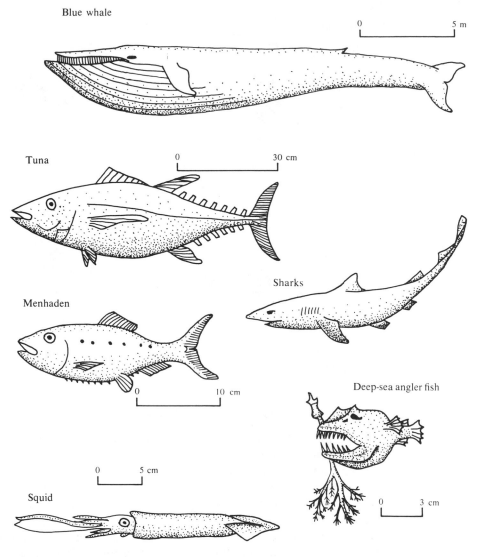

FIGURE 8.8
Some common nektonic organisms.

at night and avoid predators during daylight by staying in the dark, aphotic zone.

Neuston *live at the upper surface of the sea*. Some live in or on the water surface. Others live 10 to 20 cm below the ocean surface. Among the neuston are jellyfish and related forms, such as *Physalia* (Portuguese Man-of-War) and *Velella* (the by-the-wind-sailor), as well as less common floating snails and sargassum weed (Figure 8.9).

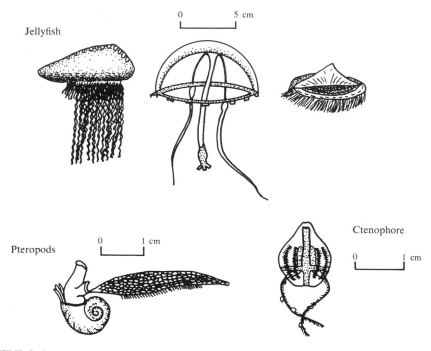

FIGURE 8.9
Some common neustonic organisms.

8.3   LIGHT AND PHOTOSYNTHESIS

Energy to support life in surface ocean waters comes from the sun. The water absorbs most incoming solar radiation, changing it to heat. Very little insolation goes into photosynthesis, typically only about 0.1%.

Even in the clearest seawater, light penetrates only the surface zone. At depths around 100 meters (about 300 ft), nearly all visible light has been absorbed. In subtropical regions where seawater is relatively free of particles and dissolved organic substances, less than 0.01% of the incoming radiation remains as visible light at 200 m. Light remaining at these depths in the open ocean is usually blue-green (Figure 8.10).

A striking feature of many open-ocean areas is the intense, almost luminous blue color of the ocean surface. The blue color shows that the waters are devoid of particles. These are oceanic "deserts" where nutrient-poor surface waters cannot support abundant phytoplankton growth. The blue color of the water is caused by the scattering and absorption of light. Pure water scatters bluish colors more readily than reddish colors. Water is also more transparent to blue light than to red.

Abundant particles in water also change water color. Colored particles or dissolved materials may themselves give seawater a variety of hues. Coastal

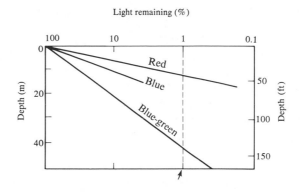

FIGURE 8.10
Amount of different colors of light remaining at various depths in the ocean, after incoming solar radiation is absorbed in near-surface waters. Note that the scale of light remaining changes by factors of ten.

waters may be brownish, greenish, or even reddish, depending on the types of sediment particles or organisms in the water. For instance, an abundance of red-colored organisms (such as dinoflagellates) causes "**red tides**," sometimes seen in coastal waters. These are patches of waters discolored by the organisms growing in them.

Suspended particles and abundant dissolved materials also limit the depth of light penetration. In turbid seawater at a 10-m depth, light levels may be comparable to those at 100 m in nonturbid seawater, where the remaining light is more yellow-green. In many tropical areas, wetlands release dissolved organic substances that color the waters brownish or blackish. Marine plants normally grow at depths where light levels exceed about 1% of the incoming solar radiation. Thus, most plant growth occurs near the ocean surface (Figure 8.1).

During photosynthesis, plants combine carbon (from carbon dioxide in seawater) and water in the presence of sunlight to form carbohydrates (sugars and starches). Phytoplankton produce most of the organic matter in the ocean. Photosynthesis and respiration can be represented as follows:

PHOTOSYNTHESIS

$$6CO_2 + 6H_2O + energy \rightleftharpoons C_6H_{12}O_6 + 6O_2(gas)$$
carbon   water   carbohydrate oxygen
dioxide

RESPIRATION

To obtain energy from food, animals essentially "burn" carbohydrates—a process called **respiration**—using oxygen and releasing carbon dioxide and water, the opposite of photosynthesis.

8.4   NUTRIENTS AND PRODUCTIVITY

In addition to light, phytoplankton obtain various dissolved substances from seawater to grow and reproduce. Many substances, such as inorganic carbon from carbon dioxide, calcium, sodium, potassium, magnesium, and sulfate are abundant everywhere in ocean waters. Other substances, known as **nutrients** (specifically nitrogen and phosphorous compounds, and silica), are scarce in seawater. In surface waters, where sunlight is abundant, nutrients are removed by plant growth. Their scarcity in surface waters thus inhibits phytoplankton growth over large parts of the ocean during summer.

After they die phytoplankton decompose and release nutrients. About 90% of cells decompose in surface waters, releasing nutrients that sustain further growth by phytoplankton. About 10% of the cells decompose only after sinking below the pycnocline where they release their constituents below the photic zone. These nutrients are then unavailable to sustain immediate further growth in the photic zone above the pycnocline, and as a result, nutrients accumulate in subsurface waters where there is too little light for photosynthesis. These return to near-surface waters through vertical water movements. *Productivity of near-surface waters over most of the ocean is controlled by the rate at which nutrient-rich subsurface waters return to the photic zone.*

Coastal ocean areas are especially productive (Figure 8.11). There, nutrients in deep subsurface waters are mixed back to the photic zone by turbulence caused by storm winds and waves and by currents flowing over irregular ocean bottom. Estuarine circulation also brings nutrients to surface waters. It retains particles and planktonic organisms and their associated nutrients in the coastal ocean.

The continued flow of subsurface waters to the surface in upwelling areas supports prolific growth of phytoplankton. This in turn supports productive fisheries where small fish (such as sardines or anchovies) feed directly on large phytoplankton (diatoms) and zooplankton.

In the open ocean, especially in temperate and tropical areas, productivity is relatively low due to nutrient depletion in surface waters. Along the equator, however, nutrients return to the surface through upwelling. Thus these equatorial upwelling areas are also highly productive of phytoplankton and fish (Figure 8.11).

8.5   SEASONAL ABUNDANCES

Availability of light and nutrients control the seasonal cycles of plankton abundance. In temperate and subpolar waters, there is too little light during winter to support abundant phytoplankton growth. During late winter or early spring, the increase in light warms the surface waters, producing a thin, less-dense surface layer. This warming keeps the phytoplankton in the photic zone (Figure 8.12) and a rapid increase in phytoplankton abundance—called a **bloom**—occurs. The temporary abundance of phytoplankton in turn supports

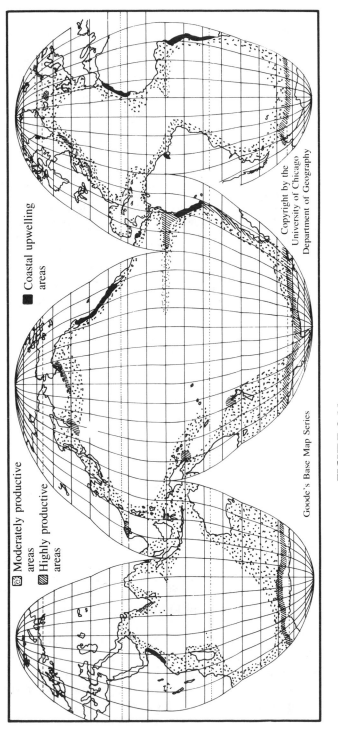

FIGURE 8.11
Distribution of productivity in the ocean.

■ Coastal upwelling
areas

▦ Moderately productive
areas

▨ Highly productive
areas

Copyright by the
University of Chicago
Department of Geography

Goode's Base Map Series

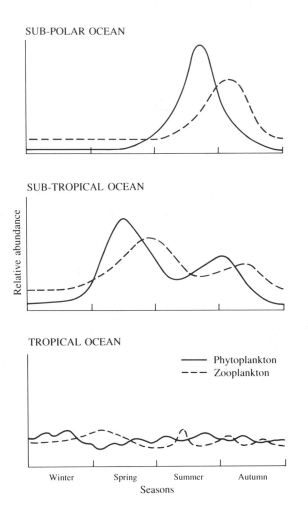

FIGURE 8.12
Patterns of phytoplankton and zooplankton abundance in various open-ocean regions.

a zooplankton bloom. The zooplankton then graze on the phytoplankton, reducing their abundance. As light intensity drops in autumn, phytoplankton produce less and eventually both phytoplankton and zooplankton occur at low levels.

Growth of phytoplankton depletes the nutrients in the near-surface waters until phytoplankton can no longer reproduce. Later, the abundance of zooplankton drops as the food supply is reduced by grazing. In autumn, storms stir the surface waters and bring more nutrients into the photic zone. If there is still enough light, other blooms occur (less intense than the spring bloom) before plankton growth drops to its wintertime level.

In tropical waters, there is always enough light to sustain photosynthesis. Thus the supply of nutrients controls phytoplankton abundance. Storms mix the waters, which causes small blooms, but none of the same intensity of the blooms that occur in higher latitude waters.

8.6 MARINE ECOSYSTEMS AND FOOD WEBS

Marine ecosystems consist of producers (plants), consumers (animals), and decomposers (bacteria, fungi). In an ecosystem, nutrients and energy from the sun are made into food by the primary producers which various organisms consume. Organisms eventually die and decompose, which releases nutrients and starts the cycle again (Figure 8.13).

Plants are **autotrophs** (self-nourishing). Using solar energy, their photosynthetic pigments form energy-rich substances such as carbohydrates. Thus *plants are the primary producers of organic matter in the ocean.* Consumers (animals) depend on plants for their food and are called **heterotrophs**. The animals which feed directly on plants are called **herbivores**. Animals that prey on other animals are called **carnivores**. **Omnivores** eat both plants and animals. After death, organisms are decomposed by **detritivores**—usually bacteria and fungi—to return nutrients back to the water.

On land, some familiar feeding relationships—called a **food chain**—are quite simple. Grass, the primary producer, is eaten by cows (herbivore), which in turn are eaten by tigers (carnivore) and humans (omnivore).

A simple oceanic feeding relationship is:

Diatoms	⟶	Copepods	⟶	Herring
100 grams		10 grams		1 gram

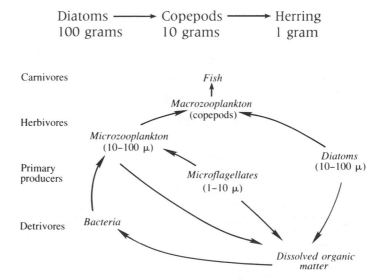

FIGURE 8.13
General food web for oceanic life.

Each stage in such a simple food chain is called a **trophic level**. Only about 10% of the energy is transferred from one trophic level to the next. Thus 100 grams of diatoms are required to produce one gram of fish at the third trophic level.

Simple food chains are rare in the ocean, especially in the open ocean. Every organism usually eats many different kinds. It in turn is eaten by a host of others. These complicated feeding relationships are called **food webs**. A relatively simple one is shown in Figure 8.14. Humans consume whales and seals from this food web. There is interest in harvesting krill as well.

8.7   MARINE MAMMALS

Marine mammals include whales, dolphins, porpoises, seals, sea lions, and walruses. They are warmblooded, breathe air, and bear their young live. All are legless and have streamlined bodies and horizontal tail flukes analogous to fish fins. They feed on other marine organisms and spend most, if not all, their life in the water. Because of their intelligence and ability to learn, dolphins, seals, and porpoises often appear in aquaria and in marine-life shows.

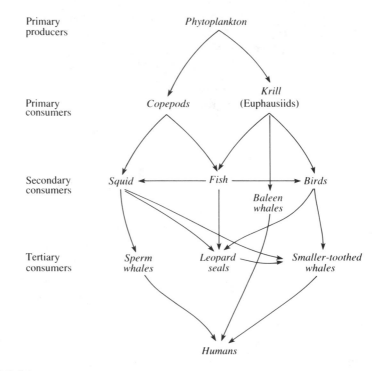

FIGURE 8.14
Simplified food web in Antarctic waters.

There are two kinds of whales. **Baleen** (whalebone) **whales** are filter feeders, have no teeth, and swim slowly, normally 3.5 km/hr (2–3 mph). They feed on planktonic crustaceans (called krill) or small fish. They filter food from the water by taking it into their mouth and then expelling it through parallel plates of hornlike baleen. Using their tongues, they scrape their food off the plates and swallow it. The largest whales, including the blue, right, hump-backed, and gray whale, are all baleen whales. Blue whales, the largest animals that ever lived, eat up to three tons of krill per day.

Toothed whales are smaller, swim faster (about 20 km/hr or 12 mph), and catch fish and squid for food. To locate their prey toothed whales use echo location systems (like the sonar systems used to locate submarines).

Because of their various feeding relationships, marine mammals occupy different positions in food webs. In Antarctic food webs (Figure 8.14), for instance, whales and seals are first-level carnivores (baleen whales, crab-eater seals), second-level carnivores (leopard seal), and third-level carnivores (toothed whales, including sperm whales).

Some of the larger whales feed in subpolar waters on the large summer plankton blooms and the bottom-dwelling organisms they support. In fall the whales migrate to warmer subtropical waters to give birth to their young. Newly born whales gain weight rapidly in preparation for the long trip back to their feeding grounds (Figure 8.15). Migrating gray whales are often seen along the U.S. Pacific coast. Toothed whales have gestation periods of more than a year, baleen whales about 11 months. Whales usually give birth to one calf at a time. A fin whale calf is about 6–7 m (20–22 ft) long at birth, about one-third the

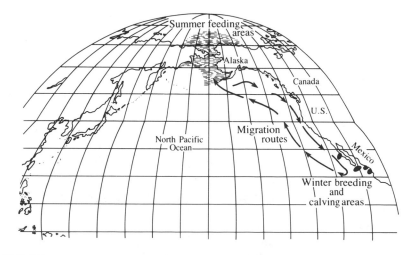

FIGURE 8.15
Migratory routes, feeding and breeding areas for California Gray Whales [From J.L. Sumich, *An Introduction to the Biology of Marine Life* 3d ed, © 1976, 1980, 1984 Wm C. Brown Publishers, Dubuque, Iowa. All Rights Reserved. Reprinted by permission]

mother's length. It suckles for about 6 months and grows to 12–13 m (37–39 ft). In 3 to 8 years, fin whales reach sexual maturity at about 20 m (65 ft).

Hunting of whales for food and oil began at least as far back as the ninth century. It was important in Europe during the Middle Ages. Whaling was important in America during the seventeenth, eighteenth, and nineteenth centuries, reaching its peak in 1850 when more than 700 whaling vessels were in operation. Whale oil was used for lamps until replaced by petroleum. Baleen was used for corset stays, among other purposes.

8.8   SEABIRDS

Seabirds are among the most conspicuous and familiar forms of marine life. Some spend nearly their entire life in the sea. Others can barely walk on land. Penguins leave the water only to lay eggs and to rear their young.

Some seabirds use the sea only for food. For example, long-billed curlews probe exposed beaches at low tide to eat animals that live deeply buried in beach sands. Short-billed shorebirds eat the shallow and smaller epifauna. Such birds are significant predators; an oyster-catcher can eat as many as 300 clams a day.

Some seabirds actively pursue their prey beneath the water. Penguins and cormorants swim underwater using either wings or feet. Pelicans fly over the water, sight a fish, then plunge from several meters above the surface to catch it.

Seabirds must drink seawater so they have a gland to secrete excess salt through their nostrils. Diving birds have a fold of skin to prevent water from entering their nostrils while diving.

To keep their feathers dry, most seabirds have an oil-secreting gland near the base of their tail. They spread the oil over their feathers by preening to waterproof them. A few birds, such as cormorants, lack enough oil, so they must spread their wings to dry in the wind after diving. Others, such as frigate birds, catch their food on the wing or by snatching it from just below the surface to avoid getting wet.

Seabirds locate food by sight, smell, and sound. They depend on finding local concentrations of food, such as schools of fish. Many seabirds live in large colonies near areas of equatorial upwelling, along upwelling coasts, or near ice margins, where productivity is high. Flocks of birds follow ships to eat garbage or animals brought up in the ship's wake.

8.9   DEEP-OCEAN-BOTTOM ORGANISMS

The types of organisms on the deep-ocean bottom and their abundance are controlled by the amount of food and by the type of bottom available—whether rocky or covered by soft sediments. Food for ocean-bottom organisms comes primarily from photosynthesis in the surface waters.

Most of the food is consumed in the photic zone and little reaches the ocean bottom, only a few percent of that produced at the surface. Much of the food reaching the bottom comes as large, rapidly sinking **fecal pellets** formed in surface waters by zooplankton. There are also infrequent falls of large animals. Because food is so scarce, large animals are rare on the deep-ocean bottom. Those living there must be able to locate infrequent food supplies. Most deep-sea animals are small and slow-growing. They reproduce infrequently, often after one of their rare meals.

Hydrothermal vents on mid-ocean ridges provide a hard bottom for attachment and an abundance of food for bottom-dwelling organisms. Because of this, active vent systems are surrounded by a rich growth of organisms—an oasis of life in an otherwise sparsely populated region. Large clams are among the many organisms living in and around vents. Clams, crabs, and fish live by feeding on the attached organisms.

Near active vents are large (up to 3 m, 10 ft long) gutless worms (Figure 8.16). Since they have no digestive tract, these animals must get their

FIGURE 8.16
Giant gutless worms, up to 2.5 m. (8 ft), grow around an active vent on the mid-ocean ridge near the Galapagos Islands, off South America. Their food comes from bacteria growing in special internal organs. The bacteria require hydrogen sulfide discharged by the vents. (Photograph by Kathleen Crane. Courtesy Woods Hole Oceanographic Institution)

food from bacteria that live inside special organs. The worms are bright red because they contain hemoglobin to transport oxygen to the bacteria that nourish them. They also have special protein systems to transport sulfides to the bacteria. Oxygen-breathing organisms that lacked these protective systems would be killed by sulfides in the water.

Volcanic activity is sporadic on mid-ocean ridges. Each large eruption apparently supports active vents with their associated organisms for a few years to a few decades. When vents are no longer active, nearby animals starve. Thus, vent organisms must be able to disperse their larvae to other active vent areas. Larvae of some of these animals are apparently planktonic for a time before returning to the bottom to attach. Others are transported by near-bottom currents.

8.10   CORAL REEFS

Coral reefs are wave-resistant structures built by carbonate-secreting organisms, mainly corals and calcareous algae. **Corals**—small, colonial animals—are the most conspicuous parts of living reefs (Figure 8.17). Encrusting **calcareous algae,** one-celled plants, coat and bind the coral skeletons, forming a large, sturdy framework. Corals and calcareous algae require sunlight. As a result, reefs grow upwards to stay near the ocean surface as sea level rises.

Coral reefs occur primarily in the tropical Indian and Pacific oceans. A few occur in the Caribbean (Figure 8.18). Most reef-building corals grow in waters with average annual temperatures between 23 and 25 °C. Most cannot tolerate prolonged exposures to cold or to large temperature changes.

There are three different types of reefs (Figure 8.19). **Fringing reefs** are connected to land and generally parallel the coast. They are largely absent near river mouths, probably killed by freshwater discharges, and usually range from a few tens to a few hundreds of meters across. **Barrier reefs** are separated by shallow lagoons from the island or mainland. In general, they parallel the coast with a few breaks large enough for ships to pass through. **Atolls** consist of a shallow lagoon surrounded by reefs with a few low, carbonate-sand islands. No volcanic island remains above sea level. This sequence of fringing reef to barrier reef to atoll was explained by Charles Darwin as being a result of the submergence of the volcanic foundation on which the corals grew. As we have already seen, the submergence is due to the general submergence of the ocean floor as it ages. Also, the sea floor subsides due to the weight of the volcano resting on it.

The **Great Barrier Reef** along the northeast coast of Australia is the world's largest reef complex. It extends about 2500 km and is about 150 km wide. More than 300 atolls are known, most in the western central Pacific and a few in the Indian Ocean (Figure 8.18).

FIGURE 8.17
Corals are the most conspicuous organisms on a reef surface. Encrusting calcareous algae cover the spaces between the coral heads. Note the abundance of fishes above the reef. (Copyright by Great Barrier Reef Marine Park Authority. Used by permission)

Reefs provide shallow-water environments where nutrients are conserved and recycled in the midst of nutrient-poor, open-ocean waters. The abundance of nutrients supports phytoplankton growth to feed benthic organisms on the reef as well as fish and plankton in the protected waters of the lagoons. In the deep ocean, coral reefs are oases of life in areas of low productivity.

Reefs are highly productive. Part of their productivity is supported by organisms producing nitrogen compounds needed by the phytoplankton. Another reason for their high productivity is the occurrence of **zooxanthella** (a type of **dinoflagellate**) within the tissues of the corals. These plants depend on the host coral for their nutrients, and in turn they provide the animal with food. In many respects, corals function as plants even though they are actually animals.

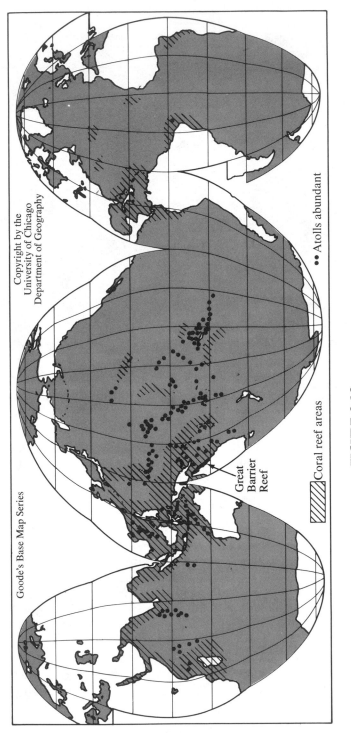

Goode's Base Map Series

Great
Barrier
Reef

Coral reef areas

•• Atolls abundant

FIGURE 8.18
Distribution of coral reefs and atolls.

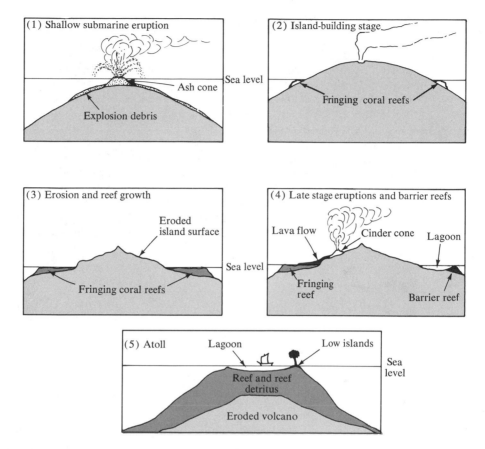

FIGURE 8.19
Stages in the evolution of atolls.

QUESTIONS

1. Define ecosystem. Describe the major functions that organisms perform in an ecosystem.
2. Discuss why upwelling conditions can support a major fishery.
3. List the types of marine organisms that are important constituents of phytoplankton.
4. Why are the deeper ocean waters richer in nutrients than the surface waters after a period of phytoplankton growth?
5. Why can many marine organisms exist without skeletons?
6. Discuss the factors controlling the abundance and distribution of organisms in the surface ocean.
7. Describe a typical grazing food chain.
9. Discuss the importance of encrusting algae in reefs.

SUPPLEMENTARY READING
Books

Burton, R. *The Life and Death of Whales.* 2d ed. London: Deutsch, 1980.

George, D. and George, J. *Marine Life: An Illustrated Encyclopedia of Invertebrates in the Sea.* New York: Wiley-Interscience, 1979.

Lockley, R. M *Ocean Wanderers: The Migratory Sea Birds of the World.* Harrisburg, PA: Stackpole Books, 1974. Elementary.

Marshall, N. B. *Ocean Life in Color.* New York: Macmillan Company, 1971. Elementary; illustrated in color.

Parsons, T. R., Takahashi, M., and Hargrave, B. *Biological Oceanographic Processes,* 3d ed. Oxford: Pergamon Press, 1984. Advanced reference.

Raymond, J. E. G. *Plankton and Productivity in the Oceans.* 2d ed. Vol. 1, *Phytoplankton.* Oxford: Pergamon Press, 1980. Standard reference.

Sumich, J. L. *Biology of Marine Life.* Dubuque, IA: William C. Brown Group, 1976. Elementary.

Articles

Isaacs, J. D. "The Nature of Oceanic Life." *Scientific American* 221(3):146–165.

KEY TERMS AND CONCEPTS

Ecosystem	Food chain	Neuston
Photosynthesis	Food web	Baleen whales
Major life zones	Zooplankton	Toothed whales
Photic zone	Trophic levels	Migrations
Phytoplankton	Benthos	Seabirds
Upwelling	Nekton	Coral reefs

9
Sediments

M ost of the ocean bottom is covered by sediment, deposits of mineral grains and rock fragments from the continents, mixed with undissolved shells and bones of marine organisms. In general, sediment deposits are thin or absent on the newly formed crust at mid-ocean ridges. Conversely, sediment deposits are thickest on the oldest crust and near continents. A large fraction of the sediment deposited in the ocean lies at the base of the continental slope, forming the continental rise (Figure 9.1). These sediment deposits are many kilometers thick, especially in marginal ocean basins and near mouths of large rivers draining areas of active mountain building.

9.1   CLASSIFICATION OF SEDIMENTS

Sediment grains are classified into three groups according to their origin and composition. Knowing the origin of the grains in a deposit permits oceanographers to work out how the deposit formed. It also helps in deciphering the history of the ocean basin and provides information about how climate has changed through Earth history.

      Lithogenous (derived from rocks) components are primarily mineral grains derived from soils that were formed on the continents and transported to the ocean by rivers, winds, or glaciers. Volcanoes also contribute lithogenous particles, such as volcanic ash. Ash from large volcanic eruptions is carried by winds

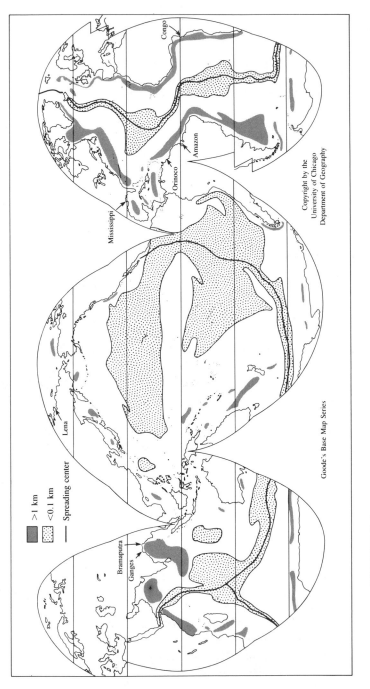

FIGURE 9.1

Thickness of unconsolidated sediments. Mouths of some major sediment-transporting rivers are shown. [After W.H. Berger Deep-sea sedimentation. In C.A. Burk and C.L. Drake (Eds.), *The Geology of Continental Margins* (Berlin: Springer-Verlag, 1974, pp. 213–242)] over many hundreds of kilometers, sometimes forming large flows that carry it down the slopes of the volcano into the sea.

Biogenous (derived from organisms) components are skeletal remains, such as shells, bones, and teeth of marine organisms. They contribute calcium carbonate, silica (opal), and phosphate minerals, forming calcareous and siliceous muds.

Hydrogenous components are formed by chemical reactions in seawater or in the sediment itself. One example is manganese nodules. We discuss each of these types of sediments in the sections that follow.

In addition to their origin and composition, sediments are also differentiated by the sizes of their grains:

> **Sand**—larger than 62 micrometers (0.062 mm) in diameter
> **Mud**—between 62 and 4 micrometers (0.004 mm)
> **Clay**— smaller than 4 micrometers.

(Remember that a human hair is 100 micrometers in diameter.)

In general, the coarsest particles in sediment deposits occur near their sources. Sands, for example, are commonly deposited on continental shelves and on beaches near the mouths of rivers. The finest grained sediments accumulate far from the sources, usually in regions of weak currents. As a result, clays are the most common deposits on the deep ocean floor (Figure 9.2). Sand-sized particles make up less than 10% of the deep-ocean sediments. The coarsest deposits are often associated with volcanic eruptions.

9.2 LITHOGENOUS SEDIMENT

Lithogenous particles are mineral grains and rock fragments derived from **weathering,** the breakdown of rocks on land. These particles are eroded by running water or ice and are carried by rivers, wind, or ice to the sea.

Small particles are carried suspended in the water. Larger ones are dragged along river bottoms by flowing waters. Sediment particles are deposited when water movements are too weak to move the grains. Wave action or currents on continental shelves sort sediment particles by size, depositing large particles and removing the smallest ones.

The largest particles remain near their source, on shores, or on beaches. Smaller particles are transported seaward and deposited in deeper water on continental shelves and slopes. The smaller particles may also be carried by the ocean surface currents or be moved by near-bottom currents. The resulting mud forms bands paralleling the coastline.

Sediment particles in water do not travel far after reaching the ocean. Most coarse-grained sediment brought to the ocean by a river accumulates in its estuary or near its mouth on a delta. Little river-borne sediment escapes the continental shelf or slope, except near major rivers, such as the Mississippi or the Ganges.

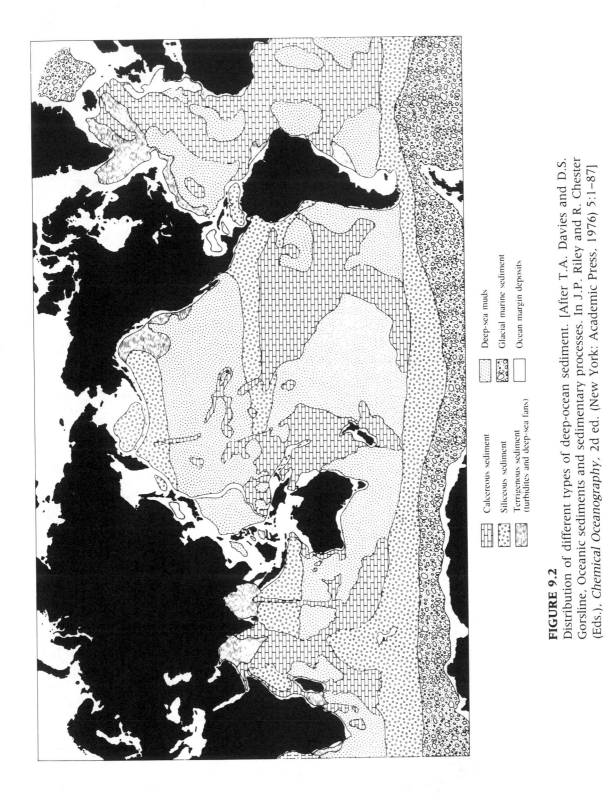

FIGURE 9.2
Distribution of different types of deep-ocean sediment. [After T.A. Davies and D.S. Gorsline, Oceanic sediments and sedimentary processes. In J.P. Riley and R. Chester (Eds.), *Chemical Oceanography*, 2d ed. (New York: Academic Press, 1976) 5:1–87]

Calcereous sediment

Siliceous sediment

Terrigenous sediment
(turbidites and deep-sea fans)

Deep-sea muds

Glacial marine sediment

Ocean margin deposits

Virtually every part of the ocean receives windblown lithogenous sediment. Far from continents, very fine-grained, deep-sea clays collect slowly, particle by particle. A layer about 1 mm thick forms in 1000 to 10,000 years (layer 1 in. thick in 25,000 to 250,000 years).

Very small, slowly sinking particles in the ocean are widely dispersed and have ample opportunity for chemical reactions to take place. For example, iron coatings on sediment particles react with dissolved oxygen in seawater, forming rust-like (iron oxide) coatings. The abundance of such red- or brown-stained grains in deep-sea clays accounts for the colors of the deposits and their common names: red clay, brown mud. Colors of deep-sea clays commonly vary from brick red in the Atlantic Ocean to chocolate brown in the Pacific. These are called **pelagic sediments** or oozes.

Near continents, and especially near large rivers, sediment deposits build rapidly, typically at rates of several meters per thousand years on continental shelves or rises. Rapidly accumulating continental shelf and slope sediments are buried too quickly to react fully with dissolved oxygen in the water. As a result, these grains do not get a reddish or brownish color. Such sediments have a variety of colors, especially green and blue colors in sediment containing large amounts of organic matter. These deposits are shown as ocean margin deposits in Figure 9.2.

Wind-transported particles are especially abundant in mid-ocean areas of the two east-west belts centered around 30°N and 30°S. High mountains and deserts are prime sources for airborne dust particles. Where the influx of lithogenous grains from rivers is low, airborne particles are a significant source of sediment. Much windblown sand from the Sahara Desert is blown out into the Atlantic.

Fine-grained volcanic ash blown into the atmosphere by volcanic eruptions settles out on the ocean surface. Ash particles are conspicuous in deep-sea sediments, accumulating in volcanic regions such as the Indonesian or Aleutian Islands.

Glaciers also supply lithogenous sediments. As glaciers flow to the coast, underlying rock surfaces are ground down into fine-grained silts by ice movements. In addition, glaciers pick up and carry rocks and boulders of various sizes. When a glacier flows into the ocean, blocks of ice break off, forming sediment-bearing icebergs. As they melt, icebergs release their sediment load. The resulting **glacial-marine sediment**—a mixture of mud, sand, and boulders—covers much of the Antarctic continental shelf (Figure 9.2).

During the most recent phase of the Ice Age, glaciers occupied large areas of the Northern Hemisphere continents, much as Antarctica is now, and nearby continental shelves were covered by glacial-marine sediment. Continental shelf areas often receive too little sediment to bury these **relict** glacial sediments. This leaves them exposed at the surface, as they are on the continental shelf off the U.S. Atlantic coast.

9.3   TURBIDITY CURRENTS

Near major sediment-transporting rivers, currents of dense, mud-rich waters—called **turbidity currents**—flow down continental slopes, carrying sediment onto the ocean floor. Such flows have never actually been observed in the ocean but are well known in lakes. Submarine cable breaks, caused by turbidity currents, are common near the mouths of major sediment-transporting rivers, such as the Congo or Ganges. Turbidity currents transport sediment originally on the continental shelf or slope and deposit it onto the deep-ocean bottom. Such deposits are called **terrigenous sediments** (Figure 9.2), meaning that they came from the land.

Although occurring infrequently, turbidity currents transport and deposit sediments over large portions of the ocean floor. For instance, a turbidity current caused by an earthquake in 1929 on the Grand Banks south of Newfoundland deposited sediment on the ocean floor over an area about 100 km long and 300 km wide. Data on the time elapsed between the earthquake and the breaking of submarine cables show that turbidity currents can travel at speeds of 20 km/hr (12 mi/hr).

Sediment-rich flows are much denser than normal seawater, causing them to flow along the ocean bottom. Turbidity currents often flow along channels on the ocean bottom, as large rivers do on land. Such flows explain the common occurrence of **turbidites**—sandy deposits having unusual textures and containing abundant shells and other remains of shallow-water organisms. Submarine canyons near the mouths of major sediment-carrying rivers would soon fill if not occasionally emptied.

Obstructions on the bottom deflect turbidity currents and can prevent their flowing onto the deep-ocean floor. This is especially obvious in the Pacific, where ridges formed by island arcs and trenches block flows of near-bottom waters toward the adjacent deep-ocean bottom. In the Atlantic, northern Indian, and Arctic oceans, turbidity currents have covered the bottom with thick sediment deposits, forming large plains.

Tops of seamounts, submarine ridges, and banks are usually not affected by turbidity currents. Sediments accumulate in these regions through particle-by-particle deposition of individual grains. The resulting deposits drape over hills rather than burying them.

9.4   BIOGENOUS SEDIMENT

Shells, bones, and teeth of marine organisms are important constituents of deep-sea sediments. Such biogenous constituents are divided into three major groups, based primarily on their chemical composition and on the organisms from which they come.

Calcareous constituents (Figure 9.3), primarily calcium carbonate, are the most abundant and are derived from shells of **foraminifera, coccoliths** (platelets secreted by tiny, one-celled algae, Coccolithophoridea), and shells of small,

(a)

(b)

FIGURE 9.3
Calcareous deep-sea sediments sampled by the *Challenger* Expedition: (a) Globigernia mud, (b) Pteropod mud.

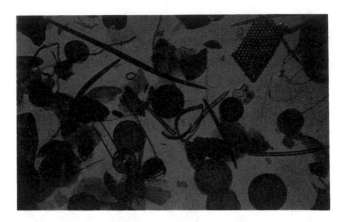

(a) Diatoms (Miocene, Spain)

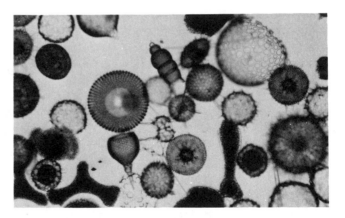

(b) Radiolaria (late Tertiary, tropical Pacific)

FIGURE 9.4
Siliceous constituents of deep-sea sediments. Diatoms (a) and radiolarians (b) are typically about the size of a sand grain (0.05 to 0.1 mm) [Photographs courtesy W.R. Reidel, Scripps Institution of Oceanography]

floating snails, the **pteropods**. Calcareous deposits (especially pteropods) occur on shallow mid-ocean ridges and oceanic plateaus (Figure 9.2).

    Siliceous constituents (Figure 9.4) are the second most abundant, being primarily opal and deriving mostly from shells of **diatoms** (one-celled **algae**) and **radiolaria** (one-celled animals). These deposits occur in equatorial and high-latitude bands (Figure 9.2).

Phosphatic constituents (rich in phosphate) are the least abundant and are derived primarily from bones, teeth, and scales of fish. These deposits are rare, occurring primarily on shallow, isolated banks or near coastal areas of high productivity.

Biogenous sediment (or ooze) is a deposit containing more than 30% biogenous constituents by volume. Since most deep-sea sediments are muds, they are called **calcareous muds** or **siliceous muds**. Calcareous muds cover nearly half the deep-ocean bottom and are most abundant on the shallower parts (less than 4500 m, or 14,700 ft). They accumulate at rates between 1 and 4 cm per thousand years.

Siliceous muds occur principally in the Pacific. Diatom-rich muds nearly surround Antarctica and occur in the North Pacific; radiolarian-rich sediment occurs in the equatorial Pacific. Both ocean areas are highly productive of marine life (Figure 9.2).

Most skeletons or shells are broken into smaller pieces or incorporated into fecal pellets. As pieces sink to the bottom, the soluble particles dissolve. The longer the grains are in contact with ocean water, the more they dissolve. **Dissolution** occurs as fragments sink or after they reach the bottom but remain unburied. Dissolution of calcareous shells is more rapid in deeper ocean waters and on the deeper ocean floor. Siliceous and phosphatic grains apparently dissolve at all depths. Consequently, certain organisms that live in profusion near the ocean surface (such as pteropods) are rarely, if ever, found in deep-sea sediment (Figure 9.5).

A final factor is **dilution**. Since biogenous sediment contains by definition more than 30% biogenous constituents, it is clear that biogenous material may not be sufficiently abundant to form a biogenous sediment in the presence of large amounts of lithogenous sediment.

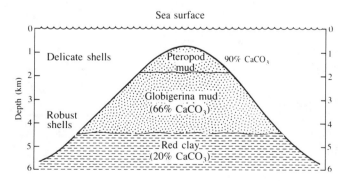

FIGURE 9.5
Idealized depth zonation of deep-sea sediments, produced by increasing dissolution of calcareous constituents. [After J. Murray and J. Hjort. *Depths of the Ocean* (London: Macmillan, 1912), p. 173]

9.5   MANGANESE NODULES

Manganese nodules are black nodules, ranging from pea- to coconut-sized. They are mixtures of iron-manganese minerals and occur on the sediment surface over much of the deep-ocean floor. Nodules cover 20 to 50% of the Pacific Ocean bottom and are the most common hydrogenous sediment in the ocean (Figure 9.6).

Nodules form very slowly, a layer a few hundredths of a millimeter to a millimeter thick forming in a thousand years. Since they accumulate extremely slowly, manganese nodules are abundant on the ocean bottom far from land and where biological productivity is low. Otherwise, lithogenous or biogenous constituents dominate the sediment. For example, the belt of hydrogenous sedi-

FIGURE 9.6
Manganese nodules are especially abundant on the deep-ocean floor in the South Pacific. These nodules are about the size of potatoes. [Courtesy IDOE. National Science Foundation]

ment is interrupted in the central Pacific by radiolarian-rich, siliceous muds along the equator (Figure 9.2). Hydrogenous constituents can form in areas of rapidly accumulating sediments, but instead of forming large nodules or slabs at the surface, they form small (pea-sized or smaller) micronodules, which are buried in the deposits.

Manganese nodules are likely commercial sources of metals. The interest in manganese nodules lies primarily in their copper, nickel, and cobalt contents. There are rich deposits in the Pacific, especially in the North Pacific, south of the Hawaiian Islands (Figure 9.7), and **cobalt-rich manganese crusts** occur at intermediate depths (2–3 km) on bare rocks around many Pacific islands. These too may have commercial uses.

9.6   BEACHES

Beaches are accumulations of loose sand and gravel moved by waves and currents. Beaches form near sources of sediment, such as eroding cliffs or river mouths. Despite their variability, beaches exhibit certain typical features. Behind a beach, there are usually **dunes** (deposits of wind-blown sand) on a low-lying shore or a sea cliff on a more rugged one. Moving toward the water, the **berm** (the backshore part of the beach) gently rises to a crest. From there the **foreshore** slopes seaward. Offshore, there are usually several submerged **bars** (Figure 9.8).

Waves dominate beach processes. Most sediment transport takes place between the upper limit of wave advance and water depths of 10 to 15 m (30 to 50 ft). Where waves are exceptionally strong, fine particles are removed, leaving coarse sand or gravel on the beach. Fine-grained materials are deposited where there is little wave action, such as in protected bays and **lagoons** behind **barrier islands**—long islands paralleling the shore—usually built of sand (Figure 9.9).

Beaches exhibit distinct seasonal cycles. In winter, beaches are eroded by the short, choppy waves caused by nearby storms. Strong currents caused by waves form deep channels, and sand removed from the beach is moved to nearby submerged bars. In summer, waves are usually low, long-period swells from distant storms, and sand moves from offshore bars to rebuild nearby beaches. Longshore bars fill in and the summer beach profile becomes less steep than during the winter.

Waves rarely approach beaches directly. Those coming in at an angle cause **longshore currents** that move sediment along the beach. These sediment movements are called **littoral drift**. Hundreds of thousands of tons of sand move each year along most beaches. Sand is trapped at inlets along barrier beaches and deposited on small submerged deltas or in lagoons behind barrier islands (Figure 9.9).

Sea level continues to rise as glaciers melt and the ocean adjusts to the disappearance of most of the continental glaciers. As sea level rises about 1.5 mm/yr (about 6 in per century), beaches move slowly landward. This causes

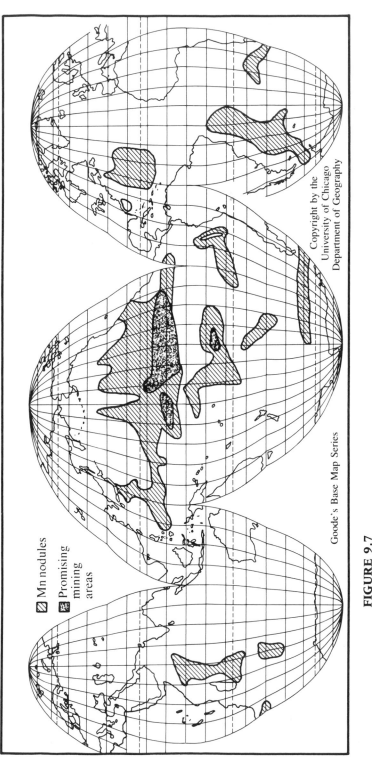

FIGURE 9.7
Distribution of rich deposits of manganese nodules and potential mining sites.

Copyright by the
University of Chicago
Department of Geography

Goode's Base Map Series

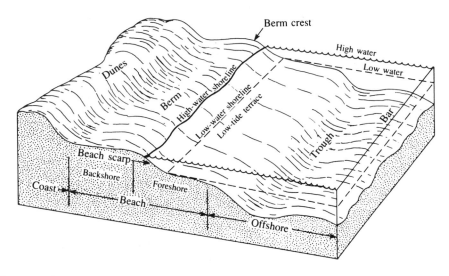

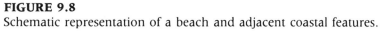

FIGURE 9.8
Schematic representation of a beach and adjacent coastal features.

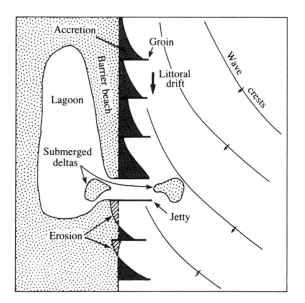

FIGURE 9.9
Waves approaching a beach obliquely cause sands to move along the beach, called littoral drift. Sediment movement is slowed and deflected by groins built to prevent beach erosion and by jetties built to protect entrances to bays.

erosion on the beachfront and sediment deposition in the lagoon and on the back of the beach. As beaches gradually move landward, previously buried lagoonal deposits are often exposed.

Since the beginning of the last glacial retreat (about 18,000 years ago), beaches have moved across the continental shelf, reworking sediment deposits. In many areas, old beach features can still be recognized. These are especially prominent where sea level has remained relatively constant for a long time.

9.7   OCEAN HISTORY

Deep-ocean sediments record the history of the overlying waters. By studying the composition of sediment layers and the fossils in them, oceanographers can decipher the ocean's history for the past 200 million years. We have already discussed the cycle of ocean basin formation and destruction in Chapter 2. Now we will consider some of the shorter-term records preserved in sediments.

Major currents such as the Gulf Stream or the Kuroshio are extremely deep, up to 2 kilometers. Where deep currents contact the bottom, they erode sediments. The eroded materials form near-bottom layers of turbid waters, which are then moved in the deep-ocean's sluggish circulation. These turbid waters occur on the western sides of the ocean basins.

Sediment deposits also record the passage of rings spun off from the major currents. These affect the bottom as **abyssal storms**. They have strong currents lasting for a few days to a few weeks and not only erode the deposits but form ripple marks like those seen in the bottoms of fast-flowing mountain streams. Areas of such ripple marks occur under the Gulf Stream off Florida and in the Carolinas, where strong currents contact the bottom.

Ripple-marked sediments also occur in areas affected by the outflows that make up the subsurface water masses. Areas off Spain and Portugal are affected by outflows from the Mediterranean. Sediments in the northwestern Atlantic record outflows of dense waters that make up the North Atlantic Deep Water.

QUESTIONS

1. What is the most common type of sediment deposit on the deep-ocean floor? Why?
2. Why are the very fine-grained sediment in deep-ocean basins commonly colored red or chocolate brown?
3. In what parts of the ocean are calcareous muds most common? Why?
4. On the average, how thick is the uncompacted sediment in the deep-ocean basins?
5. List some of the resources recovered from the ocean bottom.
6. Describe turbidity currents. List some of the evidence of their existence.
7. List three processes by which lithogenous sediment is transported into the deep ocean. Which is most important? Which is least important? Why?
8. List some of the marine organisms whose shells form biogenous sediments.
9. Why are whales' earbones and sharks' teeth found in manganese nodules?
10. What metals are likely to be recovered commercially from manganese nodules?

11. What factors make North Pacific sites most attractive for manganese nodule production? What factors work against development of North Pacific deposits?

SUPPLEMENTARY READING

Books

Kennett, J. P. *Marine Geology*. Englewood Cliffs, NJ: Prentice-Hall, Inc., 1982. Comprehensive, technical, assumes geological background.
Shepard, F. P. *The Earth Beneath the Sea*, rev. ed. Baltimore: The Johns Hopkins Press, 1967. Nontechnical.
Turekian, K. K. *Oceans*. 2d ed. Englewood Cliffs, NJ: Prentice-Hall, Inc., 1976. Elementary.

Articles

Rona, P. A. "Metal factories of the deep sea." *Natural History* 97(1):52–57.
Victory, J. J. "Metals from the Deep Sea." *Sea Frontiers* 19(1):28–33.

KEY TERMS AND CONCEPTS

Mud
Sand
Lithogenous sediment
Biogenous sediment:
   Calcareous,
   Siliceous,
   Phosphatic

Hydrogenous sediment
Red and brown muds
Manganese nodules
Glacial sediments
Turbidity currents

Metals in manganese
   nodules
Beach processes
Ocean history
Abyssal storm
Subsurface outflows

10
Using the Ocean

Increasing human populations and rising needs for foods, fuels, and other materials, mean a larger role for the ocean in our lives. This chapter discusses the factors that control development and use of ocean resources (Figure 10.1).

10.1   OCEAN RESOURCES

A natural resource is simply a supply of some material taken from our environment, such as food, water, or raw materials. **Renewable resources** are those that are replenished through growth or other processes at rates that equal or exceed rates of consumption. **Nonrenewable resources** are either not replenished or are replenished at rates much slower than the rates of consumption. Fresh water, forests, and foods are examples of renewable resources; petroleum and metals such as copper are nonrenewable. The ocean provides several renewable resources, the principal one being fresh water. Food from the sea is also a renewable resource, if properly managed.

Ocean resources are important because of the depletion of nonrenewable resources on land. The continental shelves of Texas and Louisiana have been drilled for oil and gas since the late 1940s, after new supplies on land became harder to find. Major petroleum reserves still exist on other continental shelves, including those in the North Sea and Southeast Asia. Estuaries and coastal waters have been subjected to dredging since land deposits of sand and gravel were

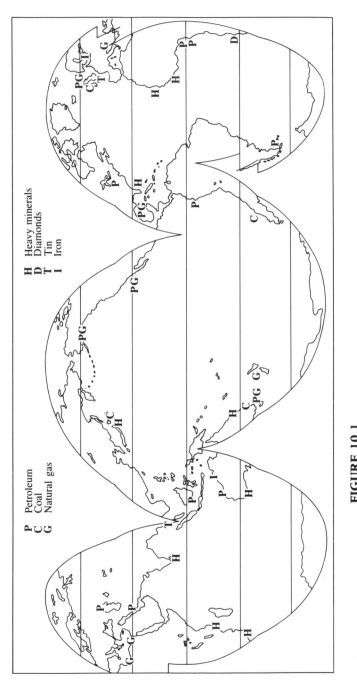

FIGURE 10.1
Fuel and mineral production from submerged deposits.

exhausted. In many cases the ocean now offers the only supply for these materials.

10.2   LAW OF THE SEA

Problems of political boundaries at sea are increasingly important as new resources are sought farther and farther from shore. In the late 1970s and early 1980s, the United Nations Conference on the Law of the Sea developed a comprehensive treaty, under which most countries claim a coastal resource zone (also called an **Exclusive Economic Zone**) extending 200 nautical mi (370 km) seaward from their shore, including offshore islands (Figure 10.2). Large areas of the ocean—about 32%—and virtually all the coastal ocean now come under the jurisdiction of a coastal state (Table 10.1). Many marginal ocean basins, for instance, the Gulf of Mexico, the Caribbean Sea, and the North Sea, are totally within the Exclusive Economic Zones of countries surrounding the ocean. Several straits—narrow ocean passages connecting adjoining ocean areas—are now within the control of coastal states.

The designation of Exclusive Economic Zones raised many questions of national boundaries. As the boundaries are extended seaward, there are many interpretations of how they should go. For the United States, this has meant boundary disputes with Canada over Georges Bank in the Atlantic and the entrance to the Strait of Juan de Fuca in the Pacific. It has also meant disputes with the USSR over Alaska and Siberia.

10.3   FISHERIES—A RENEWABLE RESOURCE

Fish provide about 3% of all protein consumed by humans and about 10% of the animal protein. For many people, especially in Asia, fish is the major protein source.

Fishing has not progressed much beyond the days of the ancients: Fish are still caught on the basis of luck and experience. While domestication of plants and animals has markedly increased our production of protein on land, we have no similar domestication in the ocean.

The world's fish production comes primarily from coastal ocean waters (Figure 10.3). Wetlands and estuaries along the coast provide breeding grounds and nursery areas for many types of fish during their critical stages of development. Coastal upwelling aids in this process by bringing nutrients to the surface zone to support rich fisheries.

Other productive fishing areas are in equatorial regions where upwelling supports abundant phytoplankton, the base of the food web that sustains fish growth. Richly productive areas also occur in Antarctic waters. Nutrient-rich waters in Antarctica occur at the surface and support luxuriant plankton growth during the brief but intense insolation of polar summers.

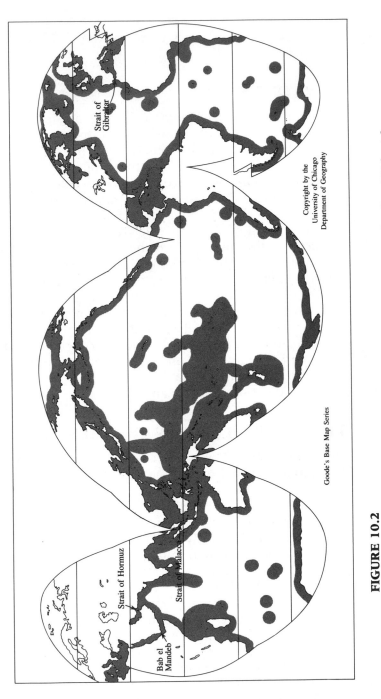

FIGURE 10.2

Exclusive Economic Zones of coastal countries extend 200 nautical miles (370 km) from their shorelines. These zones cover about one third of the ocean. Several strategic straits are also affected.

TABLE 10.1
Marine economic zones

Country	Marine Economic Area (10⁶km²)	Land Area (10⁶km²)	Ratio of Marine Economic Area to Land Area
U.S.A.	7.6	9.36	0.8
Australia	7.0	7.69	0.9
Indonesia	5.4	1.90	2.9
New Zealand	4.8	0.27	18.0
Canada	4.7	9.98	0.5
Japan	4.5	0.38	12.0
U.S.S.R.	4.4	22.4	0.2

SOURCE: Japan Marine Science and Technology Center, Yokosuka, Japan.

In 1988 the live weight of the world marine fish catch totaled 75 million metric tons. Another 7 million metric tons of various freshwater fish were caught, and about 60% of the total catch went directly into food. The remainder was either made into fish meal or oil or used for other purposes.

In 1988 the fish landed in the United States were worth $3.4 billion. Shrimp, salmon, tuna, crabs, and lobsters were the most valuable, accounting for two-thirds of the total value of the United States' catch. All of these animals

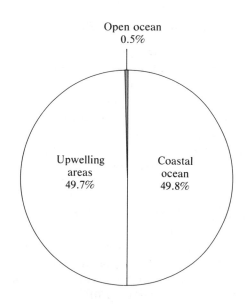

FIGURE 10.3
Most of the world's fish catch comes from the coastal ocean or from a few upwelling areas.

are coastal dwellers at some stage of their development, except tuna, which live in the open ocean.

How much fish can be produced from the ocean is hotly debated. Many scientists believe that the maximum yield might be 200 million tons a year, about three times the 1988 catch. Most agree that achieving this level requires harvesting many types of fish not now taken. Increased production might lead to serious reduction—perhaps extinction—of certain species.

10.4   MARICULTURE

One means of extracting more food from the ocean is through **mariculture** or **aquaculture** a marine equivalent of agriculture. The simplest form of aquaculture is taking animals—oysters, for example—from areas where they grow naturally and putting them in other locations where they may not spawn but can survive and fatten for market. More elaborate forms of aquaculture involve enclosures, such as fish pens, with controlled feeding and predator exclusion. In Asia, more than 3 million tons of fish are produced each year.

Lack of domesticated marine plants and animals and ignorance of fish diseases limit aquaculture. Another difficult problem concerns the uncertain legal status of many marine areas, especially in the United States. It is often impossible to control access for prevention of loss or damage to a crop. None of these problems is insurmountable, however, and mariculture is likely to become increasingly familiar as an important food source.

10.5   PETROLEUM—A NONRENEWABLE RESOURCE

Among the few substances now taken in large quantity from the ocean, petroleum and natural gas are the most familiar examples of nonrenewable resources. Petroleum is forming today, but at rates far slower than it is used.

Petroleum and natural gas are formed by decomposition of plant materials, primarily marine plants. Although most of these plants are eaten by other organisms, some of their remains are deposited in sediments. Where phytoplankton production is exceptionally high and bottom-water circulation is sluggish, dissolved-oxygen concentrations are depleted when plankton die and decay. This means that fewer benthic organisms can live in the sediment to eat organic matter; consequently, more is preserved. Eventually, some bacteria partially break down organic matter, forming gas and oil.

Eventually these sediments compact under the weight of overlying deposits, expelling water and associated oil. These fluids move through the sediments until trapped in a porous reservoir rock, typically a sandstone. After petroleum accumulates in porous and permeable rocks, it can be extracted. Thus the factors necessary for major oil and gas deposits are (1) thick accumulations of sediments rich in organic matter, and (2) permeable, porous rocks to hold petroleum in an extractable form after its movement has been stopped by (3) an impermeable layer, such as a layer of fine-grained sediment. Finally, (4) millions of years are required for these processes to occur naturally. All these factors occur on continental margins, making them attractive places to hunt for new oil and gas supplies.

Large oil and gas fields are found in the Gulf of Mexico, offshore of California and Alaska and in the Arabian Gulf. These are extensions of well-known deposits on the land. In other areas, such as the North Sea between England and Norway and in Bass Strait between Australia and Tasmania, there are no oil fields on land. All the petroleum deposits lie under coastal ocean waters.

Other promising areas for exploration include the Arctic continental shelves north of Alaska, Indonesia, and the deeper margins of North America. Future oil production may come from deposits in deeper waters on the continental rise.

10.6 RESOURCES FROM SEAWATER

Seawater itself supplies several materials: fresh water, salt, bromine, and magnesium. It is an attractive source of raw materials for several reasons. First, because of the abundance of seawater, extraction of raw materials is not likely to deplete any significant fraction of the ocean. Furthermore, the annual supply of several substances to the ocean equals or exceeds the amount extracted.

Water is, of course, the ocean's most abundant resource. Some coastal cities in arid regions derive their water from the ocean, and people are trying to duplicate the hydrologic cycle (see Chapter 3), evaporating seawater to recover fresh water. The problem is recovering the water where and when it is needed.

For centuries seawater has been an important source of salt. Evaporating basins in coastal regions use solar energy to evaporate seawater. Brine composition is controlled so only sodium chloride or other desired constituents are recovered. Normally, salt recovered from seawater is treated to remove magnesium sulfate, a bitter-tasting laxative known as Epsom salt, and calcium carbonate, a gritty impurity. Evaporating ponds operate in southern San Francisco Bay (Figure 10.4) and sea salts obtained there are used by the chemical industry, as is magnesium (Mg), a lightweight metal. Bromine (Br), also extracted from seawater, is an antiknock compound in gasoline.

Despite the wide variety of other materials dissolved in the ocean and the quantity of each substance present, the extremely dilute nature of the ocean makes it expensive to produce them. Gold is an example. It occurs in seawater at levels of about 4 grams of gold per million tons of seawater. This amounts to about 5 million tons of gold in the ocean. The cost of pumping the water and extracting the gold greatly exceeds its value.

10.7 OFFSHORE FACILITIES

Increasing populations need more space for agriculture and industry. This is especially true in extensively developed coastal regions where there are few

FIGURE 10.4
Evaporating basins are used to extract salt from seawater in southern San Francisco Bay.
(Courtesy Leslie Salt Company)

options remaining. One response to this shortage of space has been to build
new facilities on the shallow sea floor.

To the Dutch, this seaward thrust is nothing new. Between A.D. 1200 and
1950, the Netherlands reclaimed about 1.6 million acres (nearly 6300 km², or
2500 mi²). Reclamation of new areas continues. Land is reclaimed from the
shallow ocean floor by dikes that enclose fields, called **polders**. Some of these
projects have been on a monumental scale. In the 1930s, the Zuider Zee, a large
shallow embayment in the Netherlands, was cut off by a dike. This changed
it in a few years from a brackish estuary to a freshwater lake. Rich agricultural
land was reclaimed from the former bay bottom, and to prevent disastrous
flooding from storm surges much of the lower Rhine estuary was shut off from

the North Sea. Land has been reclaimed on a smaller scale in low-lying coastal areas of Great Britain.

New offshore facilities are planned to provide space for waste disposal operations (especially disposal of dredged materials), sites for large power-generating facilities, port facilities for deep-draft tankers and other vessels, and even offshore islands for industries and airports.

Offshore port facilities are used for loading and unloading supertankers, which carry hundreds of thousands of tons of petroleum and draw up to 25 to 35 m (80 to 100 ft) of water. Savings in transportation costs provided by larger vessels make them more attractive. Unfortunately, deep-draft ships cannot operate in many estuaries without extensive and expensive dredging. By locating loading or unloading facilities offshore, it is possible to avoid dredging. Offshore port facilities have been used for years in the Arabian Gulf.

There are several types of offshore facilities. The simplest are mooring and off-loading facilities for tankers. These are little more than a few pilings and moorings for the vessel to tie to. Oil moves through pipelines for storage and refining on land.

Other construction techniques suitable for building large offshore structures include: Dike and polder construction; conventional fill (including breakwater construction)—ocean bottom built up by dumping fill materials; pile supported decks—platforms built on pilings; and floating platforms—various flotation devices used to support platforms. All of these techniques have been used in various places.

10.8 ENERGY FROM THE OCEAN

For centuries, we have been harnessing the tides to produce power. In 1650 Boston had a tidally powered mill that ground corn. Modern tidal power units generate electricity. In the early 1990s two large tidal power installations were operating, one on the Rance Estuary in Brittany (France), the other on Kislaya Bay in the USSR near Murmansk.

Tidal power generation schemes use the tidal range to generate a head of water to power turbines. In other words, while going through a turbine, water must fall the distance between high and low tide (Figure 10.5) to release its potential energy. On land, dams are built on rivers to provide heads of one hundred meters or more.

Few coastal regions are suitable for tidal power generation. The largest tidal ranges in the ocean are around 15 m (50 ft), as, for example, in the Bay of Fundy, Nova Scotia. Tidal ranges exceeding 5 m (16 ft) are rare (Figure 10.6). The lowest practicable tidal range is about 5 m. Tidal ranges for most coasts are only 2 m, so there are few opportunities for tidal power plants. In addition, many potential sites are far from the urban and industrial centers that would use the power. Thus transmission costs would be high.

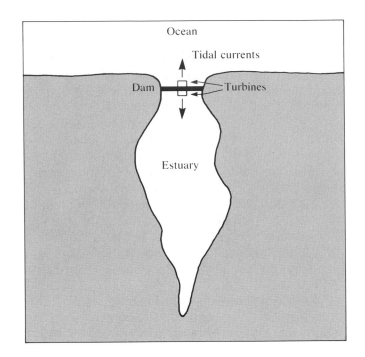

FIGURE 10.5
A tidal power plant requires a dam to enclose an inlet or estuary. Tidal currents flow through the turbines to generate power.

Topography also limits possible locations for tidal generation stations. Most tidal power plants involve one or more dams closing off the bay from the ocean. The larger and wider the opening into the bay or estuary, the more expensive the dam required. Many attractive sites are in higher latitudes, where glaciers have cut deep, narrow embayments and scoured landscapes down to bedrock.

Finally there is the problem of timing power generation. Tidal power generation is tied to the tidal cycle, which shifts by 50 minutes each day. Peak tidal power thus rarely comes during times of peak power demand. There are several schemes to alleviate this problem, one of which is to connect the station into large power distribution networks so the power is used somewhere regardless of when it is generated. Another procedure is to build several dams and to store water at high levels so one basin is a reservoir and another is a collector. This plan is expensive and calls for large, often scarce, sites. Another scheme is to store power for later use. Energy can be stored by either pumping water to high reservoirs (so that when power is needed, the water can be released to run downhill through turbines to generate power) or by making synthetic fuels, like hydrogen, which can then be stored until needed.

Waves are yet another potential energy source. A single wave 1.8 m (6 ft) high in water 9 m (30 ft) deep has about 10 kilowatts of power in each meter

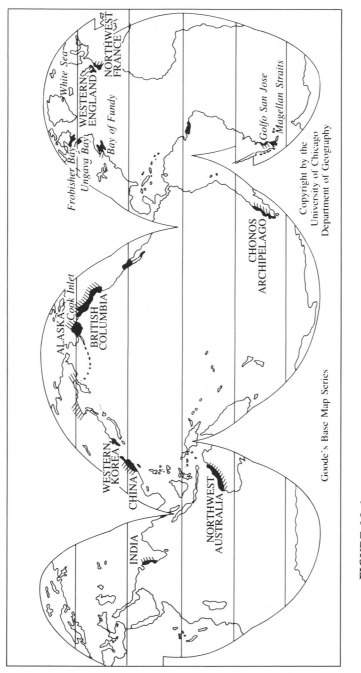

FIGURE 10.6
Potential sites for tidal power generation where average spring tidal ranges exceed 5 m (16 ft).

of wave front. The power dissipated on a beach during a single storm is enormous. Power generated by waves has been used for whistles and gongs on navigational buoys for decades. The problem is to develop ways for large-scale extraction of energy at reasonable costs.

10.9   DRUGS AND OTHER MATERIALS FROM MARINE ORGANISMS

Marine organisms contain many compounds that are potentially useful for medical purposes. For example, some corals provide compounds that kill microbes. A snail yields a compound that serves as a muscle relaxant. Even the cement that barnacles use in attaching themselves to rocks is a potential adhesive for use in cementing teeth together.

Other useful substances are also extracted from marine algae. **Agar,** a jellylike substance extracted from red algae, is used to grow bacteria in the laboratory. About one million pounds of agar is produced each year from red algae gathered from coastal waters.

Another substance is **carrageenan,** also extracted from red algae. These algae grow in the cold waters of Canada and New England, and each year about 10 million pounds of carrageenan go into medicines and foods (such as ice cream and tooth paste) as stabilizers and emulsifiers to prevent separation of the constituents.

Scientists screen plants and animals to see if they contain other useful substances. Once they isolate a substance, scientists then determine its chemical structure. The usual practice is to study the compound in the lab and then manufacture it rather than extract it from plants or animals.

10.10   EL NIÑO

Use of the ocean is affected by interactions between the ocean and atmosphere. The most striking example is the **El Niño**—the irregular (3 to 7 years) appearance of anomalously warm surface waters off Peru and Ecuador around Christmas time (Figure 10.7).

In the equatorial ocean, surface waters are quite warm. The surface layer is about 100 m thick off South America but about 300 m thick on the western side of the basin near Asia. This wedge-shaped surface zone is maintained by the prevailing Trade Winds, which blow toward the west near the equator. These winds cause upwelling off South America, bringing cold waters from below the pycnocline to the surface and through time causing a thick layer of warm waters to accumulate near Asia.

The presence of a thick lens of warm waters causes more storms than usual in the area. Eventually, something—usually a twin set of hurricanes—causes a pulse of surface waters to move eastward along the equator across the Pacific, and when this pulse of warm water reaches the coast of South America, it

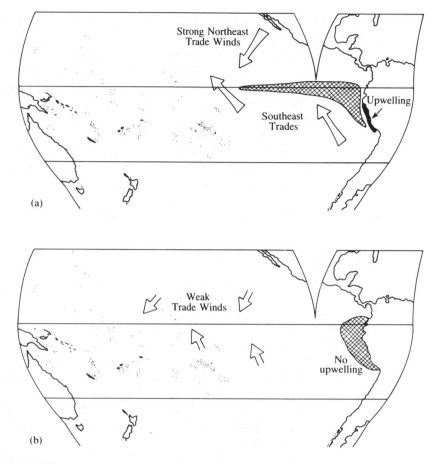

FIGURE 10.7
Conditions in the equatorial Pacific Ocean (a) before and (b) during an El Niño.

thickens the surface layer. Now the upwelled waters come from the warm, nutrient-poor surface waters rather than the cold, nutrient-rich subsurface waters. These warm surface waters cause high rainfall and flooding along the normally desert coast of South America. Elsewhere the atmospheric circulation changes, causing droughts as far away as Australia.

The sudden reduction in nutrient supply to the surface layer means that the production of food by phytoplankton in the surface zone is substantially reduced. This in turn reduces the food supply to the marine animals and fish that normally feed on them. Fishing collapses as there are fewer fish and as those that survive move offshore or deeper to find colder waters, beyond the reach of the nets used by local fishermen. Fish-eating birds starve and abandon their eggs and young as they seek to find enough food to survive themselves. El Niños in 1968–69, 1972–73, and 1982 decimated the Peruvian fishery, once the world's largest. Effects on marine life due to El Niños are felt along the coast as far away as Alaska.

El Niño has a profound influence on weather. During the exceptionally strong El Niño of 1982, the California coast was battered by unusually severe winter storms. After an El Niño, the ocean reverts to a more normal state. Cold water upwells along the equator and off the coast of South America. Sometimes, the system rebounds to an unusual extent. After the El Niño of 1987, there was a very large band of cold water along the equator. This had the effect of altering the hurricane season in the Gulf of Mexico and causing the jet stream to shift northward over North America. This is thought to have been one of the contributing factors to the North American drought of 1988, the worst since the 1930s. Understanding these connections between ocean and atmosphere will permit better long-range weather predictions.

10.11   RISING SEA LEVEL

Sea level continues to rise—about 1.5 mm per year—as glaciers melt and the ocean adjusts to the disappearance of continental glaciers. Some of the rise is due to glaciers melting and returning water to the ocean. Warming of the ocean will also increase its volume as seawater expands.

As sea level rises, beaches move slowly landward (Figure 10.8). This causes erosion on the beach front and sediment deposition in the lagoon and on the back of the beach. As beaches gradually move landward, previously buried lagoonal deposits are often exposed.

Since the beginning of the last glacial retreat (about 18,000 years ago), beaches have moved across the continental shelf, reworking sediment deposits. In many areas, old beach features can still be recognized. These are especially prominent where sea level remained relatively constant for a long time.

As Earth's climate warms, sea level rises faster. The exact amount of sea level rise is unknown just as the amount of possible global warming is unknown. The estimates of possible sea level rise with anticipated warming for the year 2100 are:

> Present rate +18 cm
> Global warming, low estimate +50 cm
> high estimate +200 cm.

Since most coasts are relatively low lying, each centimeter rise in sea level corresponds to a retreat of the shoreline by many meters. A 2-meter rise in sea level would cause substantial flooding in Florida as well as in many other coastal states. Many beach homes will be destroyed during storms as sea level rises.

To protect properties, many seawalls will be constructed. Beaches will also be replenished by supplying sand. If coarser sand than that naturally occurring is used, it is likely to remain on the beaches. It is likely, however, that the sands will be moved by littoral drift. If finer grained sands are used, they

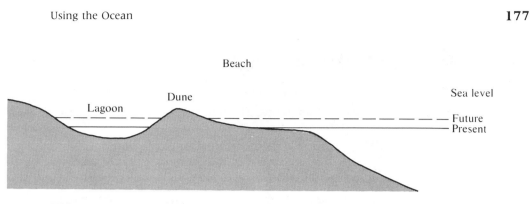

FIGURE 10.8
Rising sea level causes beaches to move landward.

are likely to be removed by wave action and deposited on the nearby ocean bottom out of the reach of waves. Such projects are costly since many thousands of cubic yards of sand are required at costs of $5 to $10 per cubic yard (1990 prices).

10.12   MARINE BIOTECHNOLOGY

No marine organisms have yet been domesticated, but efforts are underway to improve the yields of marine products through biotechnology. **Biotechnology** is the manipulation of organisms to obtain useful products or to change the characteristics of organisms themselves. This is a much faster and more controllable process than the traditional selective breeding that has been used with domesticated land animals over many thousands of years.

One technique in biotechnology is to insert a gene into a bacterium to make it produce large quantities of a substance controlled by the particular gene involved. When isolated and purified, the product can be used for many purposes. This technique has been used to make growth hormones to treat trout and striped bass. The fish can be made to eat and grow twice as fast as animals in the wild. In the future, this may be used with many different fish.

Still another approach is to inject new genetic material directly into fish eggs. Some of the genetic material is incorporated by the developing egg to make a **transgenic** animal—one that does not occur in nature. These fishes will likely first be grown in controlled enclosures for mariculture to prevent their accidental release to the environment.

QUESTIONS

1. List and discuss the factors that control development of tidal power installations. What factors would be most significant for the New York City region? For Cook Inlet, Alaska?
2. Discuss renewable and nonrenewable ocean resources. Give examples of each.
3. List some important uses of the coastal ocean. Which ones are incompatible?

SUPPLEMENTARY READING
Books

Jensen, A. C. *The Cod*. New York: Thomas Y. Crowell Company, 1972.

Laporte, Leo. *Encounter with the Earth*. San Francisco: Canfield Press, 1975. Discussion of resources, wastes and hazards.

Skinner, B. J. and Turekian, K. K. *Man and the Ocean*. Englewood Cliffs, NJ: Prentice-Hall, Inc., 1973. Elementary treatment of resources, waste disposal, and other uses of the ocean.

van Veen, J. *Dredge, Drain, Reclaim: The Art of a Nation*. 5th ed. The Hague: Martinus Nijhoff, 1962. Popular account of Dutch polder building.

Articles

Holt, S. J. ''The Food Resources of the Ocean.'' *Scientific American* 221(3):178–197.

Pinchot, G. B. ''Marine Farming.'' *Scientific American* 223(6):14–21.

KEY TERMS AND CONCEPTS

Natural resources	Mariculture	Salt
Renewable resources	Aquaculture	Distilled water
Nonrenewable resources	Productivity	Offshore facilities
Law of the Sea	Upwelling zones	Drugs from the sea
Exclusive Economic Zone	Petroleum	Energy
	Evaporation basins	El Niño
Fisheries		

Index